elementary structures for architects and builders

elementary structures for architects and builders

R. E. SHAEFFER

School of Architecture
Florida A & M University
Consulting Engineer, Tallahassee, Florida

Prentice Hall, Englewood Cliffs, New Jersey 07632

Library of Congress Cataloging-in-Publication Data

SHAEFFER, R. E.
  Elementary structures for architects and builders.

  Adaptation of: Building structures / R.E. Shaeffer.
1980.
  Includes index.
  1. Structures, Theory of. 2. Structural design.
I. Shaeffer, R. E. Building structures. II. Title.
TA645.S479 1988   624'.17   87-14538
ISBN 0-13-253014-7

Appendix H table extracted with permission from *Design Values for Wood Construction*, 1986 edition, published by the National Forest Products Association.

Appendix J table extracted with permission from the *Manual of Steel Construction*, eighth edition, published by the American Institute of Steel Construction, Inc.

Cover design: Diane Saxe
Manufacturing buyer: Peter Havens
Full-page illustrations: Pat Pinnell
Cover photograph: Dan DeForge

© 1988 by Prentice-Hall, Inc.
A Division of Simon & Schuster
Englewood Cliffs, New Jersey 07632

Printed in the United States of America

10 9 8 7 6 5 4 3 2 1

ISBN 0-13-253014-7 01

Prentice-Hall International (UK) Limited, *London*
Prentice-Hall of Australia Pty. Limited, *Sydney*
Prentice-Hall Canada Inc., *Toronto*
Prentice-Hall Hispanoamericana, S.A., *Mexico*
Prentice-Hall of India Private Limited, *New Delhi*
Prentice-Hall of Japan, Inc., *Tokyo*
Simon & Schuster Asia Pte. Ltd., *Singapore*
Editora Prentice-Hall do Brasil, Ltda., *Rio de Janeiro*

This work is dedicated to

Jane
Joseph
Kristin

contents

PREFACE   xiii

1 OVERVIEW   1

    1-1   Definition of Structure   *2*
    1-2   Structure of Buildings   *2*
    1-3   Structural Planning and Design   *5*
    1-4   Types of Loads   *6*
    1-5   Types of Stress   *7*
    1-6   Structural Forms in Nature   *7*
    1-7   Structural Forms in Buildings   *11*
    1-8   Cost   *16*
    1-9   Building Codes   *17*
    1-10   Accuracy of Computations   *17*

2 STATICS   19

    2-1   Introduction *20*
    2-2   Forces *21*
    2-3   Components and Resultants *21*
          Problems *25*
          Problems *28*
    2-4   Equilibrium of Concurrent Forces *28*
          Problems *33*
    2-5   Moments and Couples *33*
          Problems *38*
    2-6   Ideal Support Conditions *40*
    2-7   Equilibrium of Single Members *42*
          Problems *49*
    2-8   Two-Force Members *50*
          Problems *55*
    2-9   Stability and Determinacy *57*
          Problems *59*
    2-10   Simple Cable Statics *60*
          Problems *66*
    2-11   Conclusion and Procedure *68*

**3 STRUCTURAL PROPERTIES
   OF AREAS**   71

    3-1   Introduction *72*
    3-2   Centroids *72*
          Problems *78*
    3-3   Moment of Inertia *80*
          Problems *87*
    3-4   Parallel Axis Theorem *89*
          Problems *93*
    3-5   Radius of Gyration *95*
          Problems *96*

4 STRESS AND STRAIN   97

    4-1   Types of Stress *98*
          Problems *99*
    4-2   Basic Connection Stresses *101*
    4-3   Strain *102*
          Problems *103*
    4-4   Stress Versus Strain *103*
    4-5   Stiffness *105*
          Problems *106*

4-6   Total Axial Deformation   *107*
       Problems   *108*
4-7   Thermal Stresses and Strains   *108*
       Problems   *111*

5 PROPERTIES
OF STRUCTURAL MATERIALS   113

5-1   Introduction   *114*
5-2   Nature of Wood   *114*
5-3   Concrete and Reinforced Concrete   *118*
5-4   Structural Steel   *120*
5-5   Masonry and Reinforced Masonry   *121*
5-6   Creep   *121*

6 SHEAR AND MOMENT   123

6-1   Definitions and Sign Conventions   *124*
6-2   Shear and Moment Equations   *126*
       Problems   *134*
6-3   Significance of Zero Shear   *135*
       Problems   *137*
6-4   Load, Shear, and Moment Relationships   *138*
       Problems   *142*

7 FLEXURAL STRESSES   145

7-1   Introduction   *146*
7-2   Flexural Strain   *147*
7-3   Flexural Stress   *148*
       Problems   *154*
7-4   Section Modulus   *156*
       Problems   *159*
7-5   Lateral Buckling and Stability   *159*

8 SHEARING STRESSES   163

8-1   Nature of Shearing Stress   *164*
8-2   Diagonal Tension and Compression   *164*
8-3   Basic Horizontal Shearing Stress Equation   *167*
       Problems   *173*
8-4   Horizontal Shearing Stresses in Timber Beams   *174*
       Problems   *175*

8-5   Horizontal Shearing Stresses in Steel Beams   *176*
Problems   *178*

9 DEFLECTION   179

9-1   Introduction   *180*
9-2   Moment-Area Method   *181*
Problems   *188*
9-3   Principle of Superposition   *189*
Problems   *189*
9-4   Use of Deflection Formulas   *190*
Problems   *193*
9-5   Superposition and Indeterminate Structures   *193*
Problems   *196*

**10 ELASTIC BUCKLING
OF COLUMNS**   197

10-1   Columns as Building Structural Elements   *198*
10-2   Column Failure Modes   *199*
10-3   The Euler Theory   *200*
Problems   *204*
10-4   Influence of Different End Conditions   *204*
Problems   *209*
10-5   Intermediate Lateral Bracing   *210*
Problems   *213*
10-6   Limits to the Applicability of the Euler Equation   *214*

11 TRUSSES   217

11-1   Introduction   *218*
11-2   Analysis by Joint Equilibrium   *221*
Problems   *229*
11-3   Method of Sections   *230*
Problems   *234*
11-4   Special Types of Trusses   *235*

APPENDICES   239

A   Derivation of Basic Flexural Stress Equation   *240*
B   Derivation of Basic Horizontal Shearing Stress Equation   *244*
C   Derivation of Euler Column Buckling Equation   *247*
D   Weights of Selected Building Materials   *249*
E   Properties of Selected Materials   *250*

F   Properties of Areas   *251*
G   Proof of Moment-Area Theorems   *253*
H   Allowable Stress Values for Selected Woods   *257*
I   Wood Section Properties   *258*
J   Properties of Selected Steel Sections   *259*
K   Shear, Moment, and Deflection Equations   *260*

ANSWERS TO PROBLEMS   263

INDEX   277

preface

This beginning text has been written for students of architecture, building construction and the related technologies. It is intended to provide the text material for a first course in structures treating the essential topics in statics and mechanics of materials and providing an introduction to structural analysis. The presentation is basically quantitative and will be most effective when used in conjunction with a book emphasizing a qualitative approach to structural behavior.

It is assumed that the student has a background in materials and methods of construction from prior coursework or individual experience. Chapter 5 provides a very brief review of the essential characteristics of a few structural materials but it is not sufficient in depth or scope.

A minimal background in calculus and physics has been assumed in writing this material. Most of the derivations of the equations have been placed in the appendices as they are usually not absolutely essential to the use of the equations themselves. Better students, however, will gain additional understanding and insight by consulting these derivations as they are referenced.

The examples and problems are presented entirely in the customary system of units currently subscribed to in the United States. Much of the material has been extracted or adapted from *Building Structures: Elementary Analysis and Design*, a more encompassing text (Prentice-Hall, 1980), written by the author using SI Metric units exclusively.

R. E. Shaeffer

elementary structures for architects and builders

1

overview

1-1 DEFINITION OF STRUCTURE

The word "structure" has many meanings. Dictionaries usually define it in very general terms such as the following: "the organization or interrelation of all the parts of a whole; manner of construction." Structures or structured things exist almost everywhere and any definition will apply more aptly to some than others. Without confining our use of the word to buildings or other engineered objects, we find that almost everything has structure. It is very difficult to think of anything that is totally without structure. Certainly, every material object has a basic molecular structure, if nothing else. Even outer space, closer to a true vacuum than anything we know, is somewhat defined by the relatively few objects in it. It has been suggested that electrical discharge in the form of lightning has no structure. If we narrow our general terms slightly, we say that lightning seems to behave as if it has no structure. However, it has direction, and this in itself indicates the presence of some structure.

Even intangible things such as thoughts, emotions, and social relationships frequently have definite patterns. Almost everything we do or think has a structure. It is important for us to realize that in this text and our related studies, we are dealing with a very specific and narrow use of the term.

1-2 STRUCTURE OF BUILDINGS

Considering only the engineering essentials, the structure of a building can be defined as the assemblage of those parts which exist for the purpose of maintaining shape

2

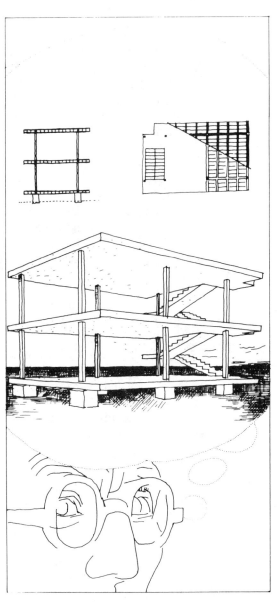

"Structure" is the assembly of parts which maintain the stability of a building. The desire to make the distinction between "working" and secondary elements, and the increasing ability to make it correctly, underlay the origins and growth of modern architecture. The primitive hut's essential structural naturalness was held up as a model of ideal clarity for architecture by the Abbe Laugier in his 1753 *Essay on Architecture*; Le Corbusier's Maisons Dom-ino of 1914, conceived as basic housing, followed Laugier's thinking by identifying slab, column, and footing as the minimum essential structure. Note that the actual structure, revealed in plan and sections, is disguised in the desire for a forceful diagram; the "slab" is a system of beams and joists.

and stability. Its primary purpose is to resist any loads applied to the building and to transmit those to the ground.

In terms of architecture, the structure of a building is and does much more than that. It is an inseparable part of the building form and to varying degrees is a generator of that form. Used skillfully, the building structure can establish or reinforce orders and rhythms among the architectural volumes and planes. It can be visually dominant or recessive. It can develop harmonies or conflicts. It can be both confining and emancipating. And, unfortunately in some cases, it cannot be ignored. It is physical.

A structural system is engineered to maintain the architectural form. Therefore, structures for buildings must be rational in terms of their adherence to the fundamental principles of science. Artists can sometimes generate shapes that seemingly ignore any consideration of natural forces, but architects cannot. The principles and tools of physics and mathematics provide the basis for differentiating between rational and irrational forms of construction.

There are at least three items that must be present in the structure of a building:

Stability
Strength and stiffness
Economy

Taking the first of the three requirements, it is obvious that *stability* is needed to maintain shape. An unstable building structure implies unbalanced forces or a lack of equilibrium and a consequent acceleration of the structure or its pieces. (The nature of structural stability is covered in more detail in Section 2-9.)

The requirement of *strength* means that the materials selected to resist the stresses generated by the loads must be adequate. Indeed, a "factor of safety" is usually provided so that under the anticipated loads, a given material is not stressed to a level even close to its rupture point. The material property called *stiffness* is considered with the requirement of strength (i.e., the structure designed must be of sufficient strength *and* stiffness). Stiffness is different from strength in that it involves how much a structure strains or deflects under load. A material that is very strong but lacking in stiffness will deform too much to be of value in resisting the forces applied.

Economy of a building structure refers to more than just the cost of the materials used. Construction economy is a complicated subject involving raw materials, fabrication, erection, and maintenance. Design and construction labor costs and the costs of energy consumption must be considered. Speed of construction and the cost of money (interest) are also factors. In most design situations, more than one structural material requires consideration. Competitive alternatives almost always exist, and the choice is seldom obvious.

Apart from these three primary requirements, several other factors are worthy of emphasis. First, the structure or structural system must *relate* to the building's function. It should not be in conflict in terms of form. For example, a linear function demands a linear structure, and therefore it would be improper to roof a bowling alley with a dome. Similarly, a theater must have large, unobstructed spans but a fine restaurant probably should not. Stated simply, the *structure* must be *appropriate* to the *function* it is to *shelter*.

Second, the structure must be *fire-resistant*. It is obvious that the structural system must be able to maintain its integrity at least until the occupants are safely out. Building codes specify the number of hours for which certain parts of a building must resist the heat without collapse. The structural materials used for those elements must be inherently fire-resistant or be adequately protected by fireproofing materials. The degree of fire resistance to be provided will depend upon a number of items, including the use and occupancy load of the space, its dimensions, and the location of the building.

Third, the structure should *integrate* well with the building's circulation systems. It should not be in conflict with the piping systems for water and waste, the ducting systems for air, or (most important) the movement of people. It is obvious that the various building systems must be coordinated as the design progresses. One can design in a sequential step-by-step manner within any one system, but the design of all of them should move in a parallel manner toward completion. Spatially, all the various parts of a building are interdependent.

Fourth, the structure must be *psychologically safe* as well as physically safe. A high-rise frame that sways considerably in the wind might not actually be dangerous but may make the building uninhabitable just the same. Lightweight floor systems that are too "bouncy" can make the users very uncomfortable. Large glass windows, uninterrupted by dividing mullions, can be quite safe but will appear very insecure to the occupant standing next to one 40 floors above the street.

Sometimes the architect must make deliberate attempts to increase the apparent strength or solidness of the structure. This apparent safety may be more important than honestly expressing the building's structure, because the untrained viewer cannot distinguish between real and perceived safety.

1-3 STRUCTURAL PLANNING AND DESIGN

The building designer needs to understand the behavior of physical structures under load. An ability to intuit or "feel" structural behavior is possessed by those having much experience involving structural analysis, both qualitative and quantitative. The consequent knowledge of how forces, stresses, and deformations build up in different materials and shapes is vital to the development of this "sense."

Beginning this study of forces (statics) and stresses and deformations (mechanics of materials) is most easily done through quantitative methods. These two subjects form the basis for all structural planning and design and are very difficult to learn in the abstract.

In most building design efforts, the initial structural planning is done by the architect. Ideally, the structural and mechanical consultants should work side by side with the architect from the conception of a project to the final days of construction. In most cases, however, the architect must make some initial assumptions about the relationships to be developed between the building form and the structural system. A solid background in structural principles and behavior is needed to make these assumptions with any reasonable degree of confidence. The shape of the structural envelope, the location of all major supporting elements, the directionality (if any) of the system, the selection of the major structural materials, and the preliminary determination of span lengths are all part of the structural planning process.

Structural design, on the other hand, is done by both the architect and the engineer. The preliminary determination of the size of major structural elements, providing a check on the rationality of previous assumptions, is done by the architect and/or the engineer. Final structural design, involving a complete analysis of all the parts and components, the working out of structural details, and the specifying of structural materials is almost always done by the structural engineer.

Of the two areas, structural planning is far more complex than structural design. It involves the previously mentioned "feeling for structure" or intuition that comes through experience. Structural design can be learned from lectures and books, but it is likely that structural planning cannot. Nevertheless, some insight and judgment can be developed from a minimal background in structural analysis and design. If possible, this should be gained from an architectural standpoint, emphasizing the relationship between the quantities and the resulting qualities wherever possible, rather than from an engineering approach.

This study of quantitative structures can be thorough enough to permit the architect to do completely the analysis for smaller projects, although such depth is not absolutely necessary. At the very least it should provide the knowledge and vocabulary necessary to work with the consulting engineer. It must be remembered that the architect receives much more education that is oriented toward creativity than does the engineer, and therefore needs to maintain control over the design. It is up to the architect to ask intelligent questions and suggest viable alternatives. If handicapped by structural ignorance, some of the design decisions will, in effect, be made by others.

1-4 TYPES OF LOADS

In general, loads that act on building structures can be divided into two groups: those due to gravitational attraction and those resulting from other natural causes and elements. Gravity loads can be further classified into two groups: live load and dead load. Building *live loads* include people and most movable objects within the structure or on top of it. Snow is a live load. So is a grand piano, a safe, or a waterbed. *Dead loads*, on the other hand, generally include the immovable objects in a building. The walls (both interior and exterior), floors, mechanical and electrical equipment, and the structural elements themselves are examples of dead loads.

Natural forces not due to gravity that act on buildings are provided by wind and earthquakes. Wind load is a lateral load that varies in intensity with the building's height, location, and shape. Earthquakes can generate very large lateral forces that impact buildings at grade level and act in a ''shaking'' manner upon the superstructure. (Hurricanes, tornadoes, and earthquakes present special design problems and local building codes often require certain types of resistive construction.) This basic text will deal primarily with gravity loading.

1-5   TYPES OF STRESS

A fundamental concept in structural analysis is that the structure as a whole and each of its elements are in a state of *equilibrium*. This means there are no unbalanced forces acting on the structure or its parts at any point. All forces counteract one another, and this results in equilibrium. When all the forces acting on a given element in the same direction are summed algebraically, the net effect is zero and there will be no acceleration. The object does respond to the forces internally, however. It is pushed or pulled and otherwise deformed; internal stresses of varying types and magnitudes accompany these deformations.

These stresses are named by their action or behavior (i.e., tension, compression, shear, and bending). *Tensile* and *compressive* stresses which act through the axis or center of mass of an object are evenly distributed over the resisting area and result in all the material fibers being stressed to like amounts. *Shearing* stresses and, more important, *bending* stresses are not uniform and usually result in a few fibers of material being deformed to their limit while others remain unstressed or nearly so. Bending is, by far, the structurally least efficient way to carry loads.

Assuming for the moment that we have a material equally strong in tension, compression, shear, and bending, it would be best to load it in tension to achieve its maximum structural capacity. Compressive forces, if applied to a long slender structure, can cause buckling as illustrated in Figure 11-2(b). Buckling always occurs under less load than would be required to fail the materials in true compression (i.e., crushing). Of course, materials are not equal in strength when loaded in different ways. Some materials have almost no tensile strength, and generalizations are very difficult to make. As explained in succeeding chapters, shearing stresses will cause tension and compression; and bending is actually a combination of shear, tension, and compression. Because of the previously mentioned uneven distribution of stress intensity, however, bending is always the most damaging load that can be applied to any structural material.

1-6   STRUCTURAL FORMS IN NATURE

Some of the most sophisticated and efficient structures are found in plants, animals, and animal houses. Through adaptation to specific environments over time, natural forms may be refined until they are nearly perfect responses to a given set of forces. Countless examples of this type of form response or form resistance, some less successful than others, may be found all around us. Only a few are cited here, to provide a representative sample. Natural forms can be very complicated in terms of structural analysis, and the reader should not be discouraged at being unable to understand them right away.

As an educational exercise, one may wish to select two or three plant or animal structures and record some preliminary thoughts or ideas about them. What forces act on them? What types of stresses are developed? What parts are strong, weak, stiff, or flexible; and why? The same forms could then be analyzed several months or a year later after completing some formal education in structures. In some cases an object that appears simple and straightforward at first glance becomes quite complex as we learn more about structural behavior.

The egg is one of the classic examples of good structure both in terms of form and material. It is a thin shell which is very strong in compression when loaded uniformly. It is doubly curved, which provides some resistance to compression buckling, a problem with all thin shells. In contrast to its strength under uniform loads, the egg is virtually defenseless against point loads. In this case, of course, the development of resistance to point loads would be totally self-defeating.

The scallop shell sketched in Figure 1-1 is considerably stronger than the egg shell. It is much more of a permanent structure and is subject to much greater loads. It is also doubly curved but is much thicker, to provide some resistance to impact loads from predators. Of greater significance, however, is the fluting or small undulations in the surface of the shell. This greatly stiffens the shell and enables it to withstand large loads without buckling. Any type of folding or ribbing of a surface (convolution) adds stiffness, and this principle is used frequently in man-made structures—from building roofs to guard rails.

The ordinary blade of grass provides an interesting example of a form that changes shape constantly over its length. Its cross section goes from a very strong and stiff tube at the bottom through a V or arc shape at midheight and finally to a very flexible flat shape at the tip. As illustrated in Figure 1-2(b), the V shape is sometimes further refined by a stiffening rib.

A blade of grass acts much like a cantilever beam sticking vertically out of the ground. It deflects when subjected to lateral loads (such as the wind) but resists

Figure 1-1 Scallop shell.

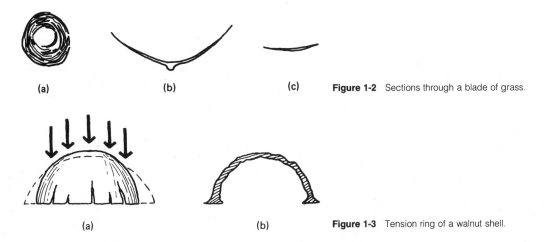

(a)   (b)   (c)   **Figure 1-2** Sections through a blade of grass.

(a)   (b)   **Figure 1-3** Tension ring of a walnut shell.

failure by having its greatest strength located at the bottom, where it is needed to resist the bending.

The common spider web is an ingenious tensile structure, light yet very strong and easily maintained. It is very redundant, possessing many extra members, and parts of it can be completely torn away and the rest will not collapse. Because every member is tensile, the structure is extremely efficient in terms of its self-weight. It is quite flexible and its highly elastic nature is well suited to the impact loads it must sustain.

One of the most sophisticated natural structures is the walnut shell. It has a double curvature and its surface has many convolutions. Figure 1-3 compares half of the shell to a dome. In general, a dome tends to thrust outward at the bottom edge as the load tries to flatten it. This circular bottom edge must be prevented from moving outward or it will develop numerous vertical cracks. One way to contain this edge is with a tension ring around the bottom. The walnut shell provides this in the form of a thickened tapered edge. This thick edge ring also helps to maintain the boundary shape against loads applied in that plane.

As if this were not enough protection for the meat inside, the interior is crossed by several paperlike tensile diaphragms which help to maintain the overall spherical shape of the shell. It is not surprising that great force must be applied to fail such a structure.

Because of the success of many natural forms, they are often copied in the design of buildings. Sometimes this is done without much thought and even less skill and the resulting design is most unfortunate. Success is more likely when we borrow the *principles* of ''resistance to loads through form'' from nature (rather than the forms themselves) and apply those principles to suit the needs of a particular design problem.

Success in the application of natural forms to architecture is more likely when the forms are regarded not as models for copying but as types of solution, demonstrations of principle, or suggestive metaphors. We are now quite accustomed to calling the structural frame a "skeleton," and to seeing the various other systems in buildings as also having purposes rather like those in animal anatomy. It was the principles on display in such places as Georges Cuvier's early 19th century museum, not the literal shapes and assemblies of bones, which inspired the analogy. The increased clarity of thinking which resulted contributed to the development of true, skeleton-frame, skyscraper construction.

1-7   STRUCTURAL FORMS IN BUILDINGS

There are several basic structural elements found in buildings, each of which embodies a different type of structural behavior. The more complicated forms are made up of combinations of the basic ones or are extensions of the same concepts. The basic elements and the stresses they develop under load are as follows:

Cable: pure tension
 Post: compression (and sometimes bending)
Beam: bending and shear[1]
Truss: tension and compression
 Arch: compression (and usually bending)
Shell: membrane (tension and/or compression evenly distributed through the shell thickness)

A considerable portion of this text is devoted to the analysis of each of these structures, except the last two. A proper discussion of arches and shells rightly belongs in a more advanced treatment of the subject, after the basic concepts are understood and some background has been established. Even so, the beginning student can develop some insight into the behavior of these and the more complicated systems from the chapters that follow.

Table 1-1 attempts to provide some data on different types of building structures. Some of these are quite conventional, while others are used only under very special circumstances. The table has been restricted to systems or parts of systems that form spans, as opposed to supporting elements, such as columns, bearing walls, or vertical cables. (This separation is somewhat arbitrary and, as seen in the table, many spanning systems act integrally with their supports.)

The tabulation is not to be considered an exhaustive classification of structural systems, and the sketches, especially, are merely representative of the class of structure listed. The figures for span range and span ratios vary widely in many cases and the values given can only be considered as guidelines. No unusual loads or support conditions have been considered in this table.

The floor systems in the flat-deck category of Table 1-1 occur more frequently than the structures of other groups by virtue of required spans and ease of construction. Each one is compatible with one or more support systems, and these relationships are shown in Figure 1-4 (see page 16).

---

[1]*Beam* is a generic term. The name applies to (in order of decreasing size and load capacity): girder, beam, joist, and purlin.

Table 1-1   Characteristics of Selected Spanning Systems

Primary means of resisting loads	*Spanning system*		*Usual materials and types*		*Usual span range (ft)*	*Typical span/depth ratio*
Tension	Cable		Steel with joist or concrete panel deck		100–500	DNA
Compression	Arch		Timber, glued-laminated		70–130	DNA
			Timber truss		100–220	DNA
			Steel truss		130–330	DNA
			Reinforced concrete, convoluted or ribbed		70–220	DNA
Bending and shear	Flat deck, floor		Wood	Joist with plywood subfloor	8–20	20
				Beam with planks	12–30	18
			Steel	Beam w/steel subfloor or concrete slab	15–50	22
				Bar joist with steel subfloor	12–60	22
			Reinf. concrete	Flat plate w/ or w/o drop panels	10–20	30
				Beam with flat slab	15–35	15
				Pan joist	15–35	20
				Waffle pan	20–50	22
				Precast plank	20–40	38
Tension and compression	Truss		Timber members		25–100	5–12
			Steel members		70–200	5–15

Typical span/ thickness ratio	Advantages	Disadvantages	Comments
300+	Long span	High technology; must provide for wind stability	Roof construction only
35	Appearance (wood finish)	Large pieces to transport	Roof construction only: usually circular or parabolic in shape
40	Low technology	Not good for concentrated loads	
40	Long span	Not good for concentrated loads	
30	Low maintenance	Slow construction	
DNA	Versatile plan and section shapes; low technology	High noise transmission	Popular for residential construction
DNA	Open walls; simple foundations; low technology	High noise transmission	Popular for residential construction
DNA	Can have composite action	Limited availability in some areas	
DNA	Wide range of available spans	No heavy loads	Deep long-span joists span much farther
DNA	Low airborne noise transmission	Short spans; openings limited	Popular for high-rise construction
DNA	Low airborne noise transmission; ease of construction	Openings limited	Square or rectangular bays
DNA	Low airborne noise transmission	Openings limited	Strong directionality
DNA	Low airborne noise transmission; appearance (concrete finish)	Poor integration w/mechanical system	Can cantilever both directions at a corner
DNA	Low airborne noise transmission; rapid construction; high span/depth ratio	Requires repetitive bay sizes	Usually prestressed
DNA	Long span	Low span/depth ratio	Popular for residential roof construction
DNA	Long span	Low span/depth ratio	Popular for industrial roof construction

Table 1-1 (*continued*)

Primary means of resisting loads	Spanning system	Usual materials and types	Usual span range (ft)	Typical span/ depth ratio
Membrane action (tension and compression)	Dome	Reinforced concrete thin shell	50–150	DNA
		Reinforced concrete (convoluted or ribbed)	100–300	DNA
		Steel truss	150–500	DNA
Membrane action (tension and compression)	Vault	Reinforced concrete thin shell	60–160	DNA
Bending and shear	Barrel vault and folded plate	Reinforced concrete thin shell	60–120	12
Tension and compression	Space frame	Steel members	70–300	30
Membrane action (tension and compression)	Warped surface	Reinforced concrete thin shell	70–200	DNA
Tension	Cable net	Steel	100–350	DNA
Membrane action (tension)	Air	Fabric-supported	150–600	DNA
		Fabric-inflated	70–180	DNA

Typical span/ thickness ratio	Advantages	Disadvantages	Comments
200	Low stresses	Slow construction; poor acoustics; no point loads	Strong directionality
40	Can accept moderate point loads	Slow construction	Popular form for sports stadia
60	Long span	Requires skin for stability	Popular form for sports stadia
175	Many shapes possible	Slow construction; openings limited; poor acoustics; no point loads	Dominant forms; usually circular or parabolic in shape; roof construction only
200	Many shapes possible	Slow construction; openings limited	Dominant forms; roof construction only; strong directionality
DNA	High span/depth ratio; long span	High technology	Roof construction only
200	Many shapes possible; low stresses	Slow construction; openings limited	Dominant forms; roof construction only
600+	Long span; rapid construction; many shapes possible	High technology; needs flexible skin system	Dominant forms; roof construction only
1000+	Long span; rapid construction; low air pressure	High noise transmission; needs air locks; fabric deterioration	Roof construction only
30	No air locks needed; rapid construction	High noise transmission; high pressure fabric deterioration	Roof construction only

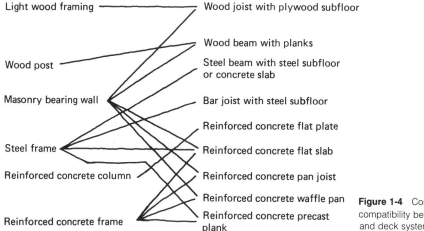

Figure 1-4 Construction compatibility between support and deck systems.

1-8 COST

Probably the question that is asked most frequently of structural consultants is one involving the relative cost of alternative structural systems. As might be suspected, it is also one of the most difficult to answer.

First, the costs of materials, fabrication, and erection are constantly changing and vary considerably with geographic location. The availability of materials and needed construction trades varies widely and transportation costs can be very high. As mentioned previously, compatibility with other building systems must also be considered (e.g., it might be costly to select a structural system that causes difficulty in the installation of mechanical ducts).

Second, and much harder to assess, is the ultimate or life-cycle cost of one system compared to another when one considers the effects that each has upon the other building systems. For example, the lowest-cost structural system might result in the greatest amount of unused building volume, which would have to be heated and cooled needlessly. The same choice could result in the highest insurance premiums. On the other hand, a low first cost might be better because the cost of money (interest) to construct the building would be less. The general subject of engineering economy treats these issues, and they will not be examined here. For the time being, it will have to be sufficient to realize that a proper determination of true cost is not simple.

On the other hand, what is not often understood by beginning designers is how low the cost of a typical building structural system, per se, really is. If we consider superstructure alone (because the cost of a foundation depends so much upon the individual site conditions), we find that it constitutes about 15 to 20% of the total construction cost. This percentage becomes even less if we consider all the costs of a project, including land, interest, fees, and overhead. The 15 to 20% range must be compared to about 35% for the mechanical and electrical systems and about 20% for the nonbearing partitions and interior finishes. These figures all vary greatly with the building type. For example, a hospital would have a high mechanical cost, driving the structural percentage down. On the other hand, a sports stadium with little interior finish costs and very long spans would have a large portion of its cost in the structure. In any event, the cost of the structural system for the great majority of buildings should probably not be a major design determinant. The beginning designer, at least, can derive greater benefits by concentrating less on structural costs and more on the relationships among structure, form, function, and space. Cost may have to compromise these relationships but should not establish them.

1-9 BUILDING CODES

Throughout this text reference is made to various codes and specifications, which provide data on design loads, allowable stresses in materials, properties and dimensions of standard cross sections, and so on. These documents also frequently spell out standard design procedures, construction tolerances, factors of safety, and in some cases even provide the appropriate design equations. Most of these are developed for the use of design professionals by the materials industry associations, which have as one of their purposes the promotion of the proper and safe use of their material.

In the United States, the large general building codes, such as that provided by the Building Officials and Code Administrators International, Inc., and the International Conference of Building Officials, and the many municipal codes often include and/or make reference to the specifications of the materials industries. In such cases these specifications, like the rest of the code, must be followed or the design professional has the responsibility of proving the equality or superiority of a different or new procedure. The intent of all building codes is the protection of the health, safety, and welfare of the public; and while some provisions may seem arbitrary or overly restrictive, they cannot be taken lightly. Indeed, the provisions of a code will generally represent minimum acceptable standards and not design ideals. As mentioned previously there are cases where the design professional has the cause and responsibility to be more stringent than the code.

A clearly written and rational building code can be of real assistance to the designer, and it is well to remember that far fewer failures occur in jurisdictions where good codes are present and enforced.

1-10 ACCURACY OF COMPUTATIONS

One of the largest inconsistencies of structural analysis and design procedures is in the determination of a proper accuracy level for computations. The author believes that sometimes it is unfortunate that we possess the machine capability to rapidly

Table 1-2 Accuracy Levels

QUANTITY	SUFFICIENT PRECISION
Member lengths	Nearest quarter foot
Cross-sectional dimension	Nearest tenth of an inch
Force	Nearest tenth of a unit (lb or kips)[a]
Moment	Nearest tenth of a unit (lb-ft or kip-ft)[a]
Stress	Nearest unit (psi or ksi)[a]
Angles	Nearest degree
Deflections	Nearest tenth of an inch
Temperature	Nearest degree

[a]Whether one works in pounds or kips depends on the magnitude of the loads and forces involved. One kip = 1000 lb.

produce answers to many decimal places. In structural engineering such precision is seldom, if ever, necessary. In most cases, any values written with more than three significant digits are misleading as to the actual accuracy level and can generate an undeserved sense of confidence.

For most purposes the levels of precision given in Table 1-2 will probably suffice and may become "too accurate" if not cut off at three significant digits. In this book the writer has attempted to round off quantities to the appropriate level except where clarity would be reduced by so doing.

2

statics

2-1 INTRODUCTION

Statics is one part of a more general subject called mechanics. *Mechanics* involves the study of forces and the effects of those forces upon the bodies on which they act. When the forces acting on a body are balanced such that no acceleration is taking place, a state of equilibrium exists. The subject of *statics* is limited to forces acting on bodies in equilibrium.

The branch of mechanics that treats unbalanced systems of forces involving acceleration is called *dynamics*. A third area, which deals with the physical deformations and internal effects in bodies caused by forces, is called *mechanics of materials* or, somewhat incorrectly, strength of materials. Mechanics of materials provides the theory behind most of the procedures used in structural analysis and design and, as such, is extensively covered in succeeding chapters of this book.

The subject matter treated in statics rarely poses any difficulty for the majority of students. Much of it is repetitious, merely repeating what has already been covered in a physics course. In some topical sequences, little emphasis is placed on statics and a thorough understanding of loads and their reactive forces is never really achieved. The reader is cautioned that statics forms the basis for all structural analysis, and all professionals in the field consider it the most important part of any study of quantitative structures.

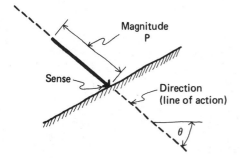

Figure 2-1 Description of a force.

2-2 FORCES

For our purposes, a *force* is a push or pull provided by one object upon another. It can act at a point, such as a concentrated column load upon a spread footing, or be distributed, such as the uniform dead load of a slab carried by a beam.

Force is a *vector* quantity, meaning that it has both magnitude and direction. If we let the magnitude of a force be represented graphically by the length of an arrow, the *direction* can be established from its line of action, as seen in Figure 2-1. The line of action extends infinitely in front of and behind the force. Each force also has a *sense*, which becomes the sign (+ or −) in algebraic computations. Sense is represented by the arrowhead in Figure 2-1. It is important not to confuse sense and direction, and it may be helpful to remember that for each direction there are two possible senses. For example, if a certain gravity load acts vertically downward, its direction is vertical and its sense is down.

In statics, we consider only external forces exerted by one body upon another. Internal forces, more properly called *stresses*, are those exerted by one part of a body upon another part of the same body. The body must be "cut open," in a figurative manner, before such forces can be examined or quantified.

2-3 COMPONENTS AND RESULTANTS

When two or more forces act at one point on a body (concurrent forces), it may be convenient to replace such forces by a single force which will have the same external effect on the body. This replacement force is called a *resultant*. Figure 2-2 shows a system of two forces being replaced by a resultant. Graphically, it is represented as the diagonal of a parallelogram, which includes the two given forces as adjacent sides. More than two concurrent forces could be treated by successive parallelograms.

It should be apparent that, by following the reverse procedure, one could resolve a given force R into two components along any two lines of action. Figure 2-3 illustrates three such sets of components. The force R could be replaced by any one of these pairs without any change in the action of the particle at O. The most

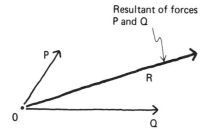

Resultant of forces
P and Q

R

0

P

Q

Figure 2-2   Resultant of two forces.

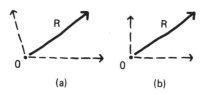

R

R

R

0   0   0

(a)   (b)   (c)

Figure 2-3   Different sets of components for
the same force R.

useful pair is that shown in Figure 2-3(b), which are called the *rectangular components*.

The force F in Figure 2-4 may be replaced by two forces, which act in the horizontal (x) and vertical (y) directions, by noting that

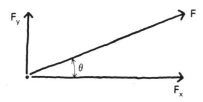

F_y

F

θ

F_x

Figure 2-4   Rectangular force components.

$$F_x = F \cos \theta \tag{2-1}$$

and

$$F_y = F \sin \theta \tag{2-2}$$

where θ is the angle made by the parent force with the x axis. Conversely, if given the two forces F_x and F_y, one could find the magnitude of their resultant by using the *Pythagorean theorem:*

$$F = \sqrt{(F_x)^2 + (F_y)^2} \tag{2-3}$$

The direction of this resultant could be found from the fact that

$$\tan\,\theta = \frac{F_y}{F_x} \tag{2-4}$$

The resultant of a system of several concurrent forces can be determined by first resolving each force into its rectangular components as shown in Figure 2-5. These components can then be treated algebraically as indicated by Equations (2-3a) and (2-4a).

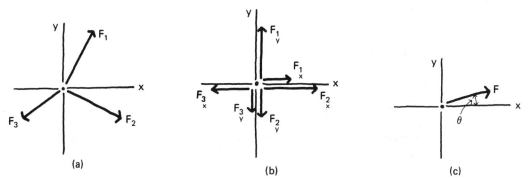

Figure 2-5

$$F = \sqrt{(\Sigma F_x)^2 + (\Sigma F_y)^2} \tag{2-3a}$$

$$\tan\,\theta = \frac{\Sigma F_y}{\Sigma F_x} \tag{2-4a}$$

Example 2-1

Determine the resultant of the three forces shown in Figure 2-6(a).

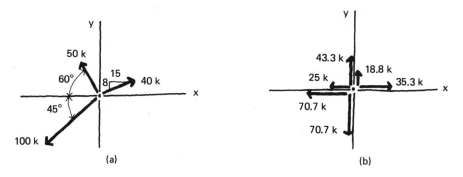

Figure 2-6

SOLUTION: Resolve each force into its rectangular components by using Equations (2-1) and (2-2). For the 50-kip force,

$$F_x = 50(0.500) = 25.0 \text{ kips}$$
$$F_y = 50(0.866) = 43.3 \text{ kips}$$

For the 40-kip force,

$$F_x = 40(\tfrac{15}{17}) = 35.3 \text{ kips}$$
$$F_y = 40(\tfrac{8}{17}) = 18.8 \text{ kips}$$

For the 100-kip force,

$$F_x = 100(0.707) = 70.7 \text{ kips}$$
$$F_y = 100(0.707) = 70.7 \text{ kips}$$

These components can then be algebraically summed in their respective directions to get the net components in Figure 2-7(a). Substituting into Equations (2-3a) and (2-4a), we obtain

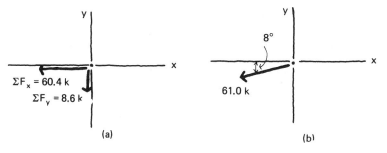

(a)   (b)   **Figure 2-7**

$$F = \sqrt{(\Sigma F_x)^2 + (\Sigma F_y)^2}$$
$$= \sqrt{(60.4)^2 + (8.6)^2}$$
$$= 61.0 \text{ kips}$$
$$\tan \theta = \frac{\Sigma F_y}{\Sigma F_x}$$
$$= \frac{8.6}{60.4}$$
$$\theta = 8°$$

PROBLEMS

2-1. Determine the magnitude, sense, and direction of the resultant of the concurrent system in Figure 2-8.

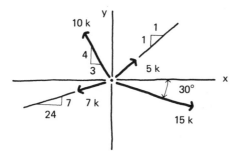

Figure 2-8

2-2. Determine the magnitude, sense, and direction of the resultant of the three forces acting at the top of the pole in Figure 2-9.

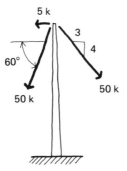

Figure 2-9

By using the fact that the opposite sides of a parallelogram are equal, we see that a force triangle (Figure 2-10) may be formed instead of the parallelogram of Figure 2.2. The components P and Q form a head-to-tail arrangement by transposing the force P. The resultant R which closes the triangle does not follow this head-to-tail order. This same procedure may be used to determine graphically the resultant of more than two concurrent forces. The forces shown in Figure 2-11(a) may be taken in any order to form a head-to-tail pattern. The force needed to close the polygon is the resultant of the system. Reversing the head-to-tail sequence will give the resultant the proper sense. Figure 2-12 shows how the rectangular components of the forces P, Q, and T will add to have the same net effect as the components of the resultant force R. Figure 2-12(d) shows two polygons (with coincident or overlapping lines),

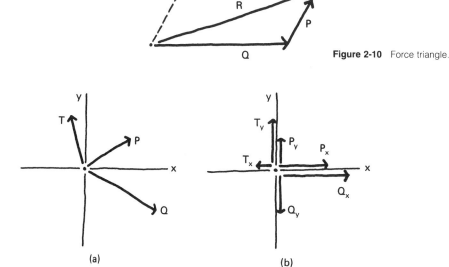

Figure 2-10 Force triangle.

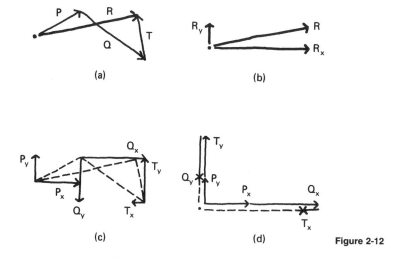

Figure 2-12

which could be formed by the rectangular components of the forces. The dashed resultants are, of course, the same as the components R_x and R_y shown in Figure 2-12(b).

The force that is equal in magnitude, but opposite in sense, to the resultant is called the *equilibrant*. It is, as its name suggests, the one force that is needed to put a given system in equilibrium. It negates or balances the effects of the other forces. Figure 2-13(b) shows the equilibrant E of the previous three-force system.

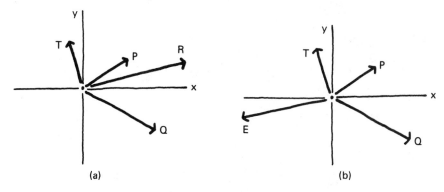

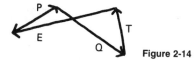

Figure 2-13   Resultant and equilibrant.

Figure 2-14

Graphically, the equilibrant will close a polygon of forces by continuing the head-to-tail relationship. In fact, each of the forces shown in Figure 2-14 is the equilibrant of the other three.

Example 2-2

Graphically determine the resultant of the force system in Example 2-1.

SOLUTION: The accuracy of any graphical solution depends upon the care and skill provided by the analyst and the scale of the drawings. Large figures will always yield a smaller percentage of error. Even so, one can seldom achieve the accuracy inherent in an algebraic solution (if such precision is necessary). In this case, the force scale is given in Figure 2-15. The answers shown in Figure 2-15 were obtained by measuring directly from the drawing with a scale and a protractor.

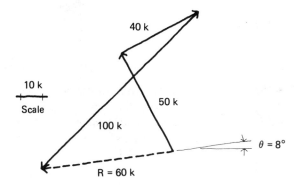

Figure 2-15   Graphical force polygon.

PROBLEMS

2-3. Work Problem 2-1 graphically.

2-4. Work Problem 2-2 graphically.

2-5. Graphically determine the equilibrant of the force system in Figure 2-16.

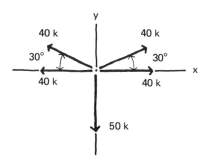

Figure 2-16

2-4 EQUILIBRIUM OF CONCURRENT FORCES

As stated in Section 2-1, equilibrium is a state of balance created by opposing forces which act on a body in such a manner that their combined net effect is zero. *Newton's first law* states, in effect, that when a body is at rest (or is moving with a constant velocity in a straight line), the resultant of the force system acting on the body is equal to zero.

If we consider, for the time being, only those force systems which are concurrent, such that there is no tendency for rotation to occur, then equilibrium can be established by setting

$$\Sigma F = 0 \qquad \qquad (2\text{-}5)$$

To ensure force equilibrium in all directions, it is usually easier to work with the two rectangular component equations:

$$\Sigma F_x = 0 \qquad \qquad (2\text{-}5a)$$
$$\Sigma F_y = 0 \qquad \qquad (2\text{-}5b)$$

(These two equations will suffice for coplanar systems; however, a third one in the z direction would be necessary for three-dimensional situations.)

The simplest structural systems are those which are concurrent and coplanar, and the two equations above will enable us to analyze such systems for two unknowns. Consider the load W suspended by the two ropes in Figure 2-17. The system is a

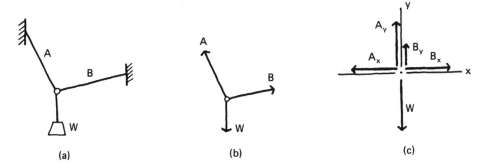

Figure 2-17

concurrent one of only three forces, A and B and the load W, as shown in Figure 2-17(b).

Assuming that the load W is known and the angles of the cables A and B are known, the forces A and B can be readily determined by using the x and y components. With reference to Figure 2-17(c) and using the standard convention that forces upward and to the right are positive in sense, we get

$$\Sigma F_x = 0 \qquad\qquad \Sigma F_y = 0$$
$$B_x - A_x = 0 \qquad B_y + A_y - W = 0$$

If the components are then expressed in terms of their parent forces A and B, the two equations can be solved simultaneously.

Problems of this type are not limited to tensile members, and the cables at A and B could just as well have been rigid bars capable of taking tension or compression. Where the sense of the force carried by such a member is uncertain, it must be assumed and then verified as the answer is obtained. A negative sign accompanying the answer will mean an incorrect sense assumption.

Example 2-3

Determine the magnitude and sense of the forces in the members A and B in Figure 2-18.

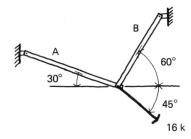

16 k   **Figure 2-18** Two bars supporting a 16-kip load.

SOLUTION: If we assume that both bars are in tension, we get the concurrent forces shown in Figure 2-19(a). Their components appear as in Figure 2-19(b). Writing the two x and y equations from the components, we get

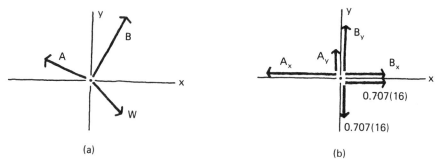

(a)   (b)

Figure 2-19

$$\Sigma F_x = 0 \qquad\qquad \Sigma F_y = 0$$
$$B_x + 0.707(16) - A_x = 0 \qquad B_y + A_y - 0.707(16) = 0$$

In terms of forces A and B,

$$0.500B - 0.866A + 11.3 = 0$$
$$0.866B + 0.500A - 11.3 = 0$$

Solving simultaneously will yield $A = +15.5$ and $B = +4.2$. The plus signs indicate that our sense assumptions were correct.

$A = 15.5$ kips tension
$B = 4.2$ kips tension

Example 2-4

Determine the magnitude and sense of the forces in bars A and B in Figure 2-20.

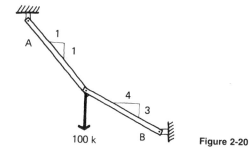

Figure 2-20

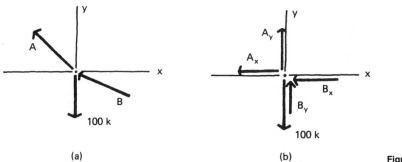

Figure 2-21

(a)   (b)

SOLUTION: For purposes of illustration, we shall assume that bar A is in tension and bar B is in compression. ~~Tension force arrows will act away from the point of concurrency and compression arrows will act in toward it~~ (Figure 2-21). (After an inspection of the horizontal force components, the reader should be able to ascertain that these sense assumptions cannot both be correct.)

$$\Sigma F_x = 0 \qquad\qquad\qquad \Sigma F_y = 0$$
$$-B_x - A_x = 0 \qquad\qquad B_y + A_y - 100 = 0$$
$$-0.800B - 0.707A = 0 \qquad 0.600B + 0.707A - 100 = 0$$

Solving for A and B, we get $A = +566$ kips and $B = -500$ kips. The minus sign indicates that bar B is actually in tension, not compression as we had assumed.

$A = 566$ kips tension
$B = 500$ kips tension

These types of problems can also be solved graphically. In each case, the three forces, load plus two unknowns, are in equilibrium. This means that each one is the equilibrant of the other two, and we must be able to form a head-to-tail triangle. The only magnitude given is that of the applied load, but the directions of all three lines of action are known. Given the three forces, for example, of Figure 2-17(b) and drawing the load vector to scale, only two possible triangles can be formed. These are shown in Figure 2-22. If we place the sense arrows on the triangles in

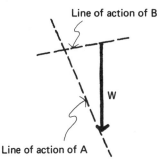

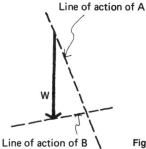

(a)   Line of action of A   (b)   Line of action of B   Figure 2-22

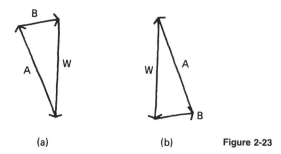

(a)   (b)   **Figure 2-23**

head-to-tail fashion as in Figure 2-23, we see that either triangle will give us the correct results. In each case, the sense arrows point away from the original point of concurrency, indicating tensile forces in both members. The magnitudes are simply measured with the force scale used to draw W.

Example 2-5

Use graphical techniques to check the results of Example 2-4.

SOLUTION: The answers in Figure 2-24(a) were obtained by measuring the polygon and are less than 5% off the values obtained in the algebraic solution. Placing the arrows so they act on the original joint, as in Figure 2-24(b), we see that both arrows represent tension.

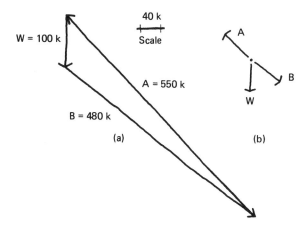

Figure 2-24 Force polygon for Example 2-5.

PROBLEMS

2-6. Determine the magnitude and sense of the forces in cables *A* and *B* of Figure 2-25.

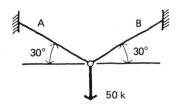

50 k   **Figure 2-25**

2-7. Determine the magnitude and sense of the forces in bars *A* and *B* of Figure 2-26. Use the algebraic method and then check your results graphically.

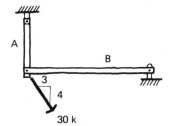

30 k   **Figure 2-26**

2-8. Determine the magnitude and sense of the forces in bars *A* and *B* in Figure 2-27.

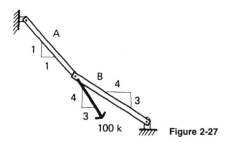

100 k   **Figure 2-27**

2-9. With reference to Figure 2-25, what will be the limiting value of the forces in bars *A* and *B* as the two angles labeled 30° approach zero?

2-5 MOMENTS AND COUPLES

A *moment* is a tendency to rotate or twist. When a force acts on a certain object, that object has a tendency to move in the direction of the force. Such motion is called

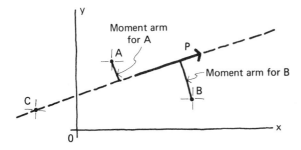

Figure 2-28   Moment of a force.

translation. Figure 2-28 shows the force P having a line of action that passes through the particle at C. Under the action of the force, the particle will tend to move along that line of action. However, the particles at A and B do not lie along the line of action. Under the influence of the force, they will not only tend to translate, but will also tend to rotate (i.e., the force P has moment with respect to those two points).[1] Indeed, a force tends to cause rotation or has moment about every point which does not lie along its line of action.

The *magnitude* of the moment of a force acting about some point is defined as the product of the force and the perpendicular distance from its line of action to the point. Such a perpendicular distance is often called a *moment arm* and is, in effect, a lever arm. The units of moment are force times distance and in structural analysis this is usually expressed in pound-feet (lb-ft) or kip-feet (kip-ft). Moment also has sense; it either acts clockwise or counterclockwise.

Most moments from forces tend to bend structural objects; for example, when two children sit on a seesaw, they bend the board, and how much it is bent depends upon the location (distance) and the weight (force) of a child. The effects of bending moments and their importance are covered in Chapters 7 and 8.

In order to obtain the magnitude of the moment of a force, with respect to

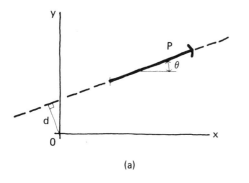

(a)

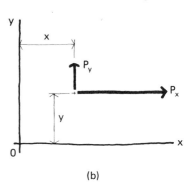

(b)

Figure 2-29   Varignon's theorem.

[1]The phrases "tend to move" or "tend to rotate" are used because in statics there are generally other forces present which act to balance out or prevent the actual motion.

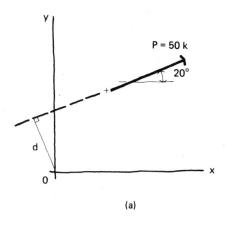

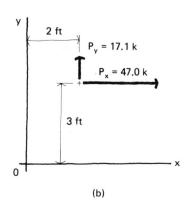

(a)   (b)

Figure 2-30

a certain point, it is sometimes easier to work with the components of the force rather than with the force itself. Figure 2-29(a) shows a force P which has a moment arm of distance d with respect to the origin O. The magnitude of this moment is P times d, and it is clockwise in sense. This value may also be found by using the components shown in Figure 2-29(b). The moment of a force with respect to a point is equal to the algebraic sum of the moments of the components of the force taken with respect to that same point. This means that the moment of P with respect to the point O can also be found as the net effect of $P_y(x)$ counterclockwise and $P_x(y)$ clockwise. [Credit for this observation is given to Pierre Varignon (1654–1722), a French mathematician.]

For algebraic purposes, we shall let counterclockwise moments be positive and clockwise moments be negative. (There is no consistency among writers in the mechanics field for this sign convention, and this selection is quite arbitrary. It is only important that one be consistent throughout a given analysis.)

To illustrate *Varignon's theorem*, let us assign values to the quantities as shown in Figure 2-30. The components in Figure 2-30(b) are found as functions of the 20° angle, and the net moment about the origin O of those components is

$$M_o = +P_y(2 \text{ ft}) - P_x(3 \text{ ft})$$
$$= +17.1 \text{ kips}(2 \text{ ft}) - 47.0 \text{ kips}(3 \text{ ft})$$
$$= -107 \text{ kip-ft}$$

or

$$M_o = 107 \text{ kip-ft} \circlearrowright$$

Alternatively, we can obtain the same value by determining the perpendicular moment arm distance d. Figure 2-31 illustrates one possible way to quantify this distance by

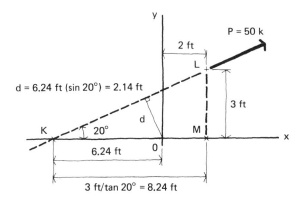

Figure 2-31 Determining the moment arm.

first using the triangle *KLM*. With *d* established as 2.14 ft, the moment of *P* about the origin is

$$M_o = -P(d)$$
$$= -50 \text{ kips}(2.14 \text{ ft})$$
$$= -107 \text{ kip-ft}$$

or

$$M_o = 107 \text{ kip-ft } \circlearrowleft$$

The reader should notice that the moment of the force *P*, being a function of the perpendicular distance to the line of action, is independent of the location of the force along its line of action. For example, if we relocate *P* so that it acts at the point *K* in Figure 2-31, then using Varignon's theorem, we get

$$M_o = P_x(0 \text{ ft}) - P_y(6.24 \text{ ft})$$
$$= -17.1 \text{ kips}(6.24 \text{ ft})$$
$$= -107 \text{ kip-ft}$$

or

$$M_o = 107 \text{ kip-ft } \circlearrowleft$$

This technique of moving a force along its line of action until one of its components passes through the desired moment center can be a valuable shortcut for many moment calculations.

In statics, we frequently encounter a special kind of tendency to rotate, provided by a force system called a couple. A *couple* consists of a pair of equal and opposite parallel forces, two forces that are equal in magnitude, opposite in sense,

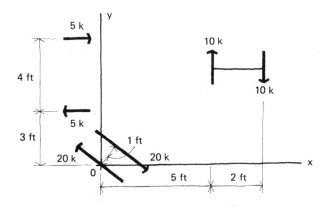

Figure 2-32 Identical couples.

and have parallel lines of action. It produces only rotation, no translation. It is pure moment. Its magnitude is given by the product of one of the forces and the perpendicular distance between the forces and its sense is clockwise or counterclockwise. The most important characteristic of a couple is that its effect (magnitude and sense of the moment) is the same with respect to every point in its plane. Its tendency to cause rotation of a point is a constant and is *independent* of the distance between the couple and the point. The student should verify arithmetically that each couple in Figure 2-32 has the same 20 kip-ft clockwise effect upon the point *O*. Given the dimensions to any other point in the plane of Figure 2-32, we could show that the same 20 kip-ft rotation acts there also.

From the standpoint of static equilibrium, the effect of a couple can only be negated or balanced by a second couple of identical magnitude but opposite sense. The equilibrant couple must lie in the same (or parallel) plane. The moment of a couple cannot be balanced by a single force because the value of moment for a force is dependent upon the location of the moment center. Similarly, the action of a single force cannot be negated by a couple, for a couple provides no translation.

An understanding of what a couple is enables us to look more closely at what happens inside a bending member (beam). In Chapter 1, mention was made of the inefficiency of bending as a means to carry load. Not only are the stresses distributed very unevenly within a beam, but the small structural depth of most beams necessitates large couple forces to maintain equilibrium. Figure 2-33 shows a beam with a single concentrated load carried between two bearing walls. The beam itself

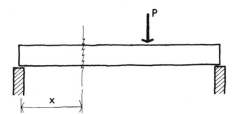

Figure 2-33 Simple beam.

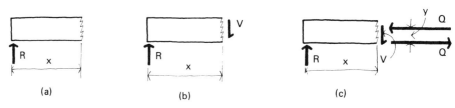

Figure 2-34

is assumed to be weightless. Let us cut through the beam near the load to examine the internal forces. Isolating the left-hand portion of the beam from the rest of the system, we will see the upward reaction R provided by the wall as in Figure 2-34(a). An internal downward force called V in Figure 2-34(b) develops as a response to R. This force V is provided by the other part of the beam and, for vertical equilibrium, would have to be exactly equal to R.

These two forces constitute a couple of magnitude V times x (or R times x), which can only be balanced by an opposing couple. In other words, the beam portion might be in force (translational) equilibrium ($\Sigma F = 0$), but it is not in moment (rotational) equilibrium. The second or opposing couple can only be provided by the other beam part upon the cut face as shown in Figure 2-34(c). Because the moment arm of this opposing couple is limited by the beam depth, the forces Q will be quite large. For moment equilibrium ($\Sigma M = 0$), Q times y must equal V times x. The two forces Q, of course, are equal according to the definition of a couple. This also assures force equilibrium in the horizontal direction.

The preceding explanation introduced the third equation of static equilibrium, which joins the previous two discussed in Section 2-4 dealing with concurrent forces. Since a beam involves forces that are not concurrent, all three equations were involved.

$$\Sigma F_x = 0 \qquad\qquad (2\text{-}5a)$$
$$\Sigma F_y = 0 \qquad\qquad (2\text{-}5b)$$
$$\Sigma M = 0 \qquad\qquad (2\text{-}6)$$

These are the only three equations of statics that are applicable to coplanar structures.

PROBLEMS

2-10. Determine the magnitude and sense of the moment of the force in Figure 2-35 with respect to points O, A, and B.

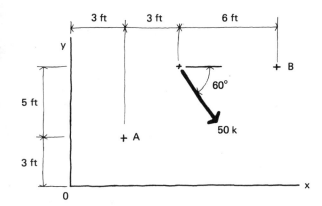

Figure 2-35

2-11. Show that the four forces in Figure 2-36 have a net moment of zero with respect to points O and A and a third point of your choice.

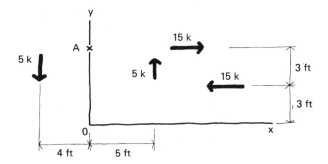

Figure 2-36

Sometimes it is computationally convenient to replace a force with a force and a couple, which have the same effect upon the member. Equilibrium is maintained if we proportion the new force and couple correctly. For example, Figure 2-37(a) shows an eccentric load acting on the top of a pier. The effect on the pier consists of a downward force (P) and a clockwise moment (P times e). In Figure 2-37(b), two collinear forces P (shown dashed) have been added to the pier. These forces cancel each other and we have, in effect, added zero to the system. However, the two forces P which are circled constitute a clockwise couple of magnitude P times e. The replacement system is shown in Figure 2-37(c) with the couple represented as a moment arc of value P(e). This new concentric force and couple have the same action upon the pier as did the original eccentric load. Almost all problems involving eccentric loads can be conceptually simplified using such equivalent systems.

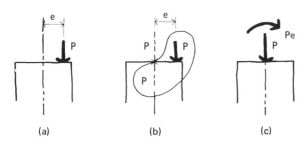

(a)   (b)   (c)   **Figure 2-37**

2-6 IDEAL SUPPORT CONDITIONS

The actual support conditions and member connections in real building structures are quite complicated. The proper design of connections, if all the various forces were considered, would be a lengthy and exacting process. An accurate determination of the amount of friction and slip, for example, which occurs under load at a given beam to column joint would be almost impossible. Nevertheless, the behavior of the supports can have a critical effect upon the structure proper, and the designer cannot ignore the end conditions of any member.

In actual practice, it is necessary to make certain idealized simplifications regarding the nature of connections and supports. The designer must be aware of the fact that these simplifications are false and may or may not approximate the actual construction conditions. Judgment must be used to increase the factors of safety involved in the design whenever it is suspected that an assumed condition departs markedly from the actual one. It is too easy to forget that, while it is simple to draw frictionless rollers or fully rigid, inflexible connections, they are literally impossible to fabricate.

The symbols presented in Figures 2-38 through 2-41 for various ideal support conditions are standard and universally accepted. What is not universally accepted is the determination of characteristics needed, by actual field connections and conditions, for the symbols to be valid representations. In spite of this, some attempt has been made in the following discussion to indicate a few types of real connections which would correspond to the various symbols.

The *hanger* can take no compression and is assumed to provide a single force of known sense and direction. The *rod* or *angles* are assumed to be long and slender, having no resistance to compression buckling and negligible bending resistance. If the tension member is placed at an angle, its force can be represented as two rectangular components, but these components are *not independent*. They are related to each other by the direction of their resultant, which acts along the member.

The *pin* can take one force of unknown sense and direction (Figure 2-39). In theory, it offers no resistance to rotation (moment). The single force is usually represented by two rectangular components, since this will cover any possible line

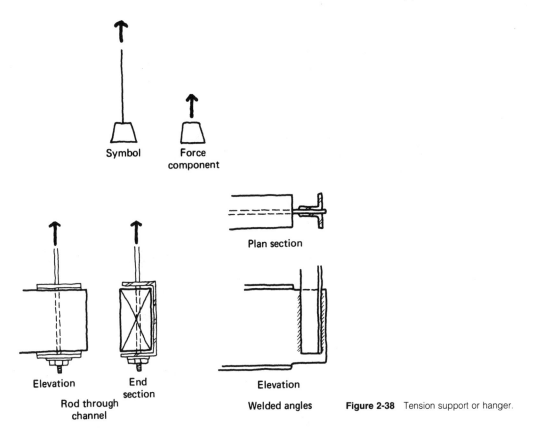

Symbol   Force component

Plan section

Elevation   End section

Rod through channel

Elevation

Welded angles

Figure 2-38 Tension support or hanger.

of action. The two components are independent as to both magnitude and sense and will constitute two unknowns.

The *roller* provides a single force of known direction (Figure 2-40). Its sense is unknown (i.e., it is assumed that the member cannot "lift off" a roller). Like the tension hanger, if the roller or link is on an angle, its force is usually represented as two rectangular components. These two components are *dependent* both as to magnitude and sense and constitute but one unknown. Once either component is determined, the other is known by trigonometry. The pin and roller are called *simple supports*.

The *link support* is really more of a separate structural member than a support, but it acts very much like a roller. To provide a force of known direction, the link must have a pin (or hinge) at both ends and carry no load in between. Sometimes called a *strut*, it is a special case of a larger group of structural elements, described later (Section 2-8) as two-force members.

The *fixed* or *rigid* or *built-in support* can take one force of unknown sense and direction (Figure 2-41). In theory, it also has full moment resistance (of either sense) and will not change its angle of attachment. Like the pin, the single force is

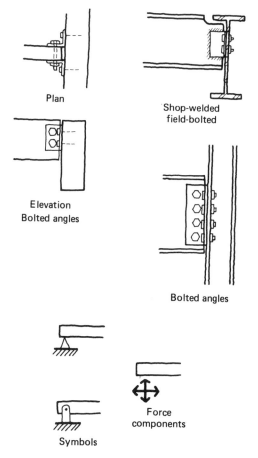

Plan

Shop-welded
field-bolted

Elevation
Bolted angles

Bolted angles

Force
components

Symbols

Figure 2-39   Pin or hinged supports.

usually represented as two completely independent components. A fixed connection has the potential of three unknowns (i.e., two independent components and a couple). This couple is sometimes called a *moment reaction* or *fixed-end moment*.

2-7 EQUILIBRIUM OF SINGLE MEMBERS

Before looking further into the subject of static equilibrium, the concept of the *rigid body* should be introduced. In statics, it becomes convenient, if not necessary, to ignore the small deformations and displacements which take place when a member is loaded. To do this, we pretend that the materials used for all structural elements

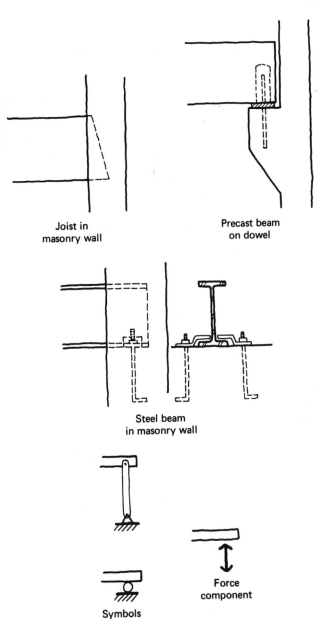

Figure 2-40   Roller and link supports.

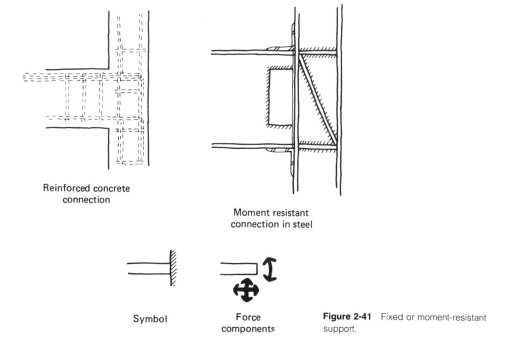

Reinforced concrete
connection

Moment resistant
connection in steel

Symbol

Force
components

Figure 2-41   Fixed or moment-resistant
support.

and supports are rigid, having the property of infinite stiffness. We assume that members do not stretch, compress, or bend in any way and that their geometry, therefore, remains constant. This is, of course, never true. Even though structural materials are very stiff, they all deform slightly even under small loads. However, the assumption that structural bodies are rigid greatly simplifies many situations in terms of static equilibrium and, in most cases, introduces an insignificant amount of error. An example of the type of minute change, which is generally ignored, is the shortening of span that takes place when a beam deflects into an arc. Depending upon the type of loading, minor changes in the upward reactions at the supports would also occur. Not only would such changes be insignificant if expressed in percentage terms, but would also be difficult to consider quantitatively because, like other deformations, they vary with the load.

Once all of the external forces have been resolved in terms of statics, the stresses and strains within the various elements of the structure are examined. At this point, it is critical that material deformations are not ignored. The rigid-body concept is useful only for the determination of external forces. (It is generally valid but can require modification when applied to the statics of more complicated structural problems.)

While the idea of a structure that is rigid for some analytical operations but not for others may seem incongruous to the novice, a little experience will quickly provide the rationale and judgment behind this concept.

In structural analysis, the principles of statics are used to determine reactive forces, which are responses to the applied loads. These reactions always develop the appropriate magnitudes and directions, such that the end result is one of equilibrium. In other words, under the combined action of the loads and reactions, each element of the structure has zero tendency to translate and zero tendency to rotate.

Determining the needed reactive forces is made easier if the analyst makes a sketch of the structure or element, showing all the forces involved (known and unknown). Such a sketch is called a *free-body diagram* (FBD), and most structural designers consider making such a diagram the first step in any statics problem. A free-body diagram shows the body in isolation or cut "free" from everything adjacent to it. The effects of all such removed objects are shown as forces acting at the appropriate locations. We have already used these diagrams in this chapter without calling them by name. Figure 2-17(b) is, in effect, a free-body diagram of the central joint. The portion of beam shown in Figure 2-34(c) is another free-body diagram. The examples that follow will all utilize an FBD.

Example 2-6

Determine the reaction components for the simply supported beam in Figure 2-42.

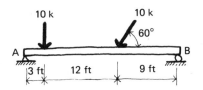

Figure 2-42 Beam with two loads.

SOLUTION: The free-body diagram is shown in Figure 2-43. The senses of the unknown reactions must be assumed. (A negative sign on the answer will mean an incorrect sense assumption.) The load that acts at an angle has been resolved into its rectangular components. The three equations of equilibrium are then used to find the unknown force components.

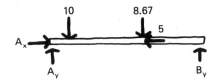

Figure 2-43 Free-body diagram for Example 2-6.

(Whenever practicable, moment equations are written for more than one point, and one of the two force equations is left for use as an independent check on the answers. It should be noted that regardless of the number of moment centers selected, however, there remain only three independent equations in planar statics, and thus a maximum of three independent unknowns can be determined. Other techniques must be used to supplement these three equations whenever the number of independent reaction components exceeds three.)

$$\Sigma F_x = 0$$
$$A_x - 5 = 0 \qquad A_x = 5 \text{ kips} \atop \rightarrow$$

$$\Sigma M_A = 0$$
$$B_y(24) - 8.67(15) - 10(3) = 0 \qquad B_y = 6.67 \text{ kips} \uparrow$$

$$\Sigma M_B = 0$$
$$-A_y(24) + 10(21) + 8.67(9) = 0 \qquad A_y = 12.0 \text{ kips} \uparrow$$

$$\Sigma F_y = 0 \quad \text{(check)}$$
$$-10 - 8.67 + 12.0 + 6.67 = 0 \quad \checkmark$$

Notice that each answer specifies magnitude, direction, and sense of the force.

Example 2-7

Determine the reaction components for the structure in Figure 2-44(a).

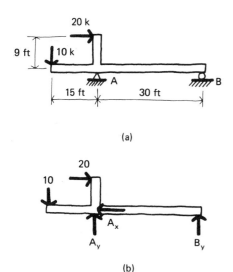

(a)

(b)

Figure 2-44   T-frame and free-body diagram.

SOLUTION:

$$\Sigma F_x = 0$$
$$20 - A_x = 0 \qquad A_x = 20 \text{ kips}$$
$$\underset{\leftarrow}{}$$
$$\Sigma M_A = 0$$
$$10(15) - 20(9) + B_y(30) = 0 \qquad B_y = 1 \text{ kip } \uparrow$$
$$\Sigma M_B = 0$$
$$10(45) - 20(9) - A_y(30) = 0 \qquad A_y = 9 \text{ kips } \uparrow$$
$$\Sigma F_y = 0 \quad \text{(check)}$$
$$-10 + 9 + 1 = 0 \quad \checkmark$$

Example 2-8

Determine the reaction components for the beam in Figure 2-45(a).

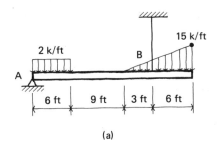

(a)

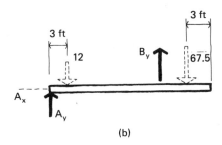

(b)

Figure 2-45 Overhanging beam and free-body diagram.

SOLUTION: The uniform load and the uniformly varying load are both converted to equivalent concentrated loads for the purpose of obtaining reactions. Each load shown dashed in Figure 2-45(b) is placed at the centroid of its load area. (See Appendix F for the triangular load.)

$$\Sigma F_x = 0 \qquad A_x = 0$$

$$\Sigma M_A = 0$$

$$B_y(18) - 12(3) - 67.5(21) = 0 \qquad B_y = 80.75 \text{ kips } \uparrow$$

$$\Sigma M_B = 0$$

$$-A_y(18) + 12(15) - 67.5(3) = 0$$

$$A_y = -1.25 \qquad A_y = 1.25 \text{ kips } \downarrow$$

(The negative sign means that the sense of A_y shown on the FBD is incorrect.)

$$\Sigma F_y = 0 \quad \text{(check)}$$

$$-12 - 67.5 + 80.75 - 1.25 = 0 \quad \checkmark$$

Example 2-9

Determine the reaction components for the frame in Figure 2-46(a).

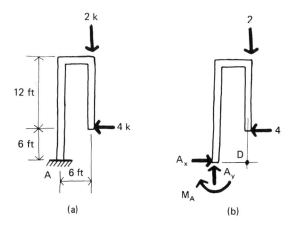

(a)   (b)

Figure 2-46 Cantilevered frame.

SOLUTION:

$$\Sigma F_x = 0$$

$$A_x - 4 = 0 \qquad A_x = 4 \text{ kips } \rightarrow$$

$$\Sigma F_y = 0$$

$$A_y - 2 = 0 \qquad A_y = 2 \text{ kips } \uparrow$$

$$\Sigma M_A = 0$$

$$-M_A - 2(6) + 4(6) = 0 \qquad M_A = 12 \text{ kip-ft } \circlearrowleft$$

To accomplish a check, select a convenient moment center such as point D.

$$\Sigma M_D = 0 \quad \text{(check)}$$
$$-M_A - A_y (6) + 4(6) = 0$$
$$-12 - 2(6) + 4(6) = 0 \quad \checkmark$$

PROBLEMS

2-12. Determine the reaction components for the simple beam of Figure 2-47.

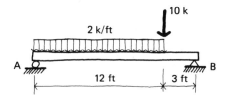

Figure 2-47

2-13. Determine the reaction components for the U-shaped frame of Figure 2-48.

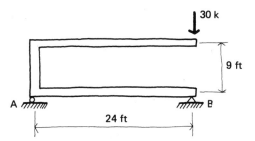

Figure 2-48

2-14. Determine the reaction components for the frame loaded by wind in Figure 2-49.

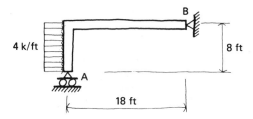

Figure 2-49

2-15. Determine the reaction components for the beam in Figure 2-50. (*Hint*: Separate the load into two parts.)

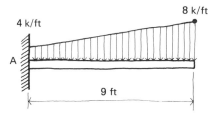

Figure 2-50   Cantilever beam.

2-16. Determine the reaction components for the beams of Figure 2-51. The connection at *B* may be assumed to act like a pin in both beams. (*Hint*: The components of the sloped cable force are related to each other by the angle of the cable.)

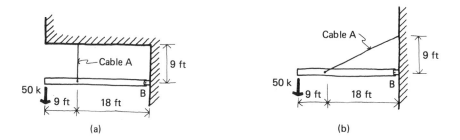

Figure 2-51   Beams suspended by cables.

2-8 TWO-FORCE MEMBERS

When a structural element is hinged or pinned at each end and carries no load in between, it is called a *two-force member*. Such elements have only two forces acting on them, one applied at each of the two pins. To maintain equilibrium, these forces must be equal, opposite, and collinear. If forces are resolved into components, the components are dependent upon the line of action of the parent force. In this case, the line of action must connect the two pins. In other words, the direction of the two forces is known by inspection.

Member *DB* in Figure 2-52 is a two-force member. Provided that no intermediate load is placed on it, any force acting at *D* or *B* must have the line of action shown. The magnitude and sense of such forces will be a function of the loading on member *AC*. There are only two possibilities for equilibrium, as shown in Figure

2-53(a) and (b). Any components drawn must be consistent as to sense, as in Figure 2-53(c) or (d). This means that if one of the four components is determined, the others are known automatically. Likewise, if one of the four components is assumed as to sense, the other three sense assumptions are made without choice.

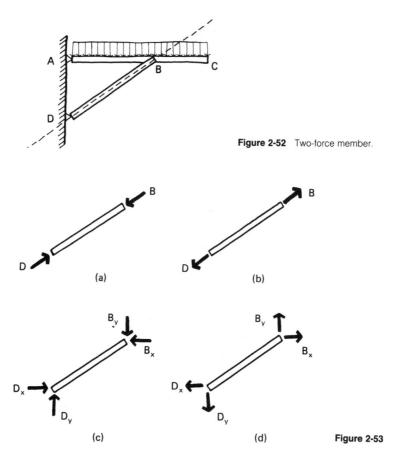

Figure 2-52 Two-force member.

Figure 2-53

A *cable* is a two-force member of limited sense (i.e., only tension). A roller is the simplest of all two-force members and the link support is by definition a two-force member. The determination of force direction by two-force members is a very useful tool in statics. It means, for example, that the reactions of the three structures of Figure 2-54 will be identical. In each case, the reaction at B (which passes through C) will be as shown in Figure 2-55(a). For algebraic determination, it is probably easiest to deal with the dependent components shown in Figure 2-55(b). The components in this case will have the relationship

$$\frac{B_x}{B_y} = \frac{3}{4}$$

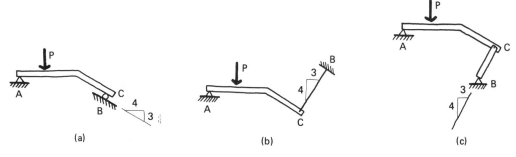

Figure 2-54   Three two-force support elements.

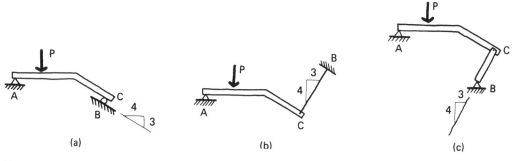

Figure 2-55   Free-body diagrams.

Because the equilibrium determination of a two-force member is trivial, it is best to remove it from free-body diagrams just as if it were a support. This was done in the examples and problems of Section 2-4 for the equilibrium of concurrent forces. Each bar or cable functioned as a two-force member and was removed when we sketched an FBD of the point of concurrency.

Example 2-10

Determine the reaction components at *A* and *B* for the structure in Figure 2-56.

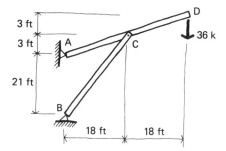

Figure 2-56

SOLUTION: The member BC is a two-force member which holds up the bar AD. (In this case, it is fairly obvious that BC acts in compression, and the force components will be assumed accordingly. When the sense is not so obvious, it may be guessed as either tension or compression, consistent with the options of Figure 2-53.) The reader should recognize from Figure 2-57(b) that $B_x = C_x$ and $B_y = C_y$. When writing the equations of equilibrium for the FBD of member AD, these equivalencies will be used. Also recognize that the two-force member is on a slope of 24 ft vertical on 18 ft horizontal, and therefore $B_x = \frac{3}{4}B_y$ or $B_y = \frac{4}{3}B_x$.

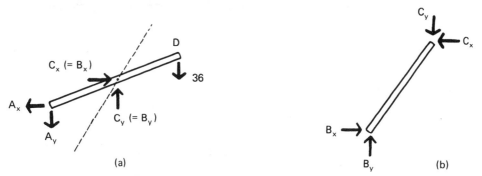

(a)   **(b)**

Figure 2-57   Free-body diagrams for Example 2-10.

$$\Sigma M_A = 0$$

$$-B_x(3) + B_y(18) - 36(36) = 0$$

$$\text{But } B_x = \frac{3}{4}B_y$$

$$-\frac{3}{4}B_y(3) + 18B_y - 1296 = 0 \qquad\qquad B_y = 82.3 \text{ kips } \uparrow$$

If B_y is 82.3, then

$$B_x = \frac{3}{4}(82.3) \qquad B_x = 61.7 \text{ kips}$$
$$\longrightarrow$$

The two-force equations can be used to get A_x and A_y from an FBD of member AD or an FBD of the whole structure, as in Figure 2-58.

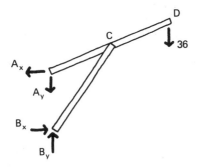

Figure 2-58   FBD of whole structure.

$$\Sigma F_x = 0$$
$$-A_x + 61.7 = 0 \qquad A_x = 61.7 \text{ kips}$$
$$\leftarrow$$

$$\Sigma F_y = 0$$
$$-A_y + 82.3 - 36 = 0 \qquad A_y = 46.3 \text{ kips} \downarrow$$

A check on our work is essential because, in this case, all the answers were made dependent upon the first answer for B_y. Using Figure 2-58 and taking moments about B, we get

$$\Sigma M_B = 0 \quad \text{(check)}$$
$$A_x(21) - 36(36) = 0$$
$$61.7(21) - 1296 \approx 0 \quad \checkmark$$

Example 2-11

Determine the reaction components at A and B for the structure in Figure 2-59(a).

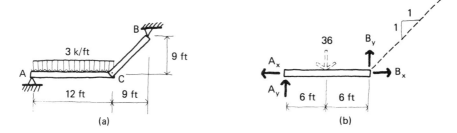

Figure 2-59

SOLUTION: The two-force member BC acts on a 45° slope, which means that $B_x = B_y$. Using the free-body diagram in Figure 2-59(b),

$$\Sigma M_A = 0$$
$$-36(6) + B_y(12) = 0 \qquad B_y = 18 \text{ kips} \uparrow$$

From the slope of BC,
$$B_x = B_y \qquad B_x = 18 \text{ kips}$$
$$\rightarrow$$

$$\Sigma M_B = 0$$
$$-A_y(12) + 36(6) = 0 \qquad A_y = 18 \text{ kips} \uparrow$$

$$\Sigma F_x = 0$$
$$-A_x + B_x = 0 \qquad A_x = 18 \text{ kips}$$
$$\leftarrow$$

Since $B_x = 18$,

$$\Sigma F_y = 0 \quad (\text{check})$$
$$+18 - 36 + 18 = 0 \quad \checkmark$$

From Example 2-11, one can see that the length of the two-force member does not affect the external forces involved; only the slope (or geometry) is important. The line of action of the forces acting on a two-force member always connects the two pins present at each end of the member. It then becomes apparent that the shape of the two-force member between the pins is arbitrary. The shape of the two-force member will *not* affect the statics of the structure. This means that the three structures shown in Figure 2-60 would have the same external reactions as we found in Example 2-11.

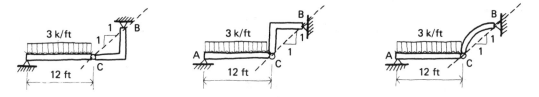

Figure 2-60   Three structures that have the same reactions as the structure in Figure 2-59(a).

PROBLEMS

2-17.   Determine the reaction components for the structure in Figure 2-61.

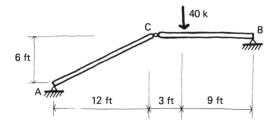

Figure 2-61

2-18.   Determine the reaction components for the structure in Figure 2-62. (*Hint:* Make an FBD of member *BC* to prove that B_y and C_y are both zero.)

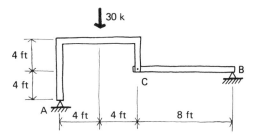

Figure 2-62

2-19. With reference to the structure in Figure 2-62, move the load so that it acts in the middle of member *BC*, and then determine the reaction components.

2-20. Determine the reactions for the structure in Figure 2-63.

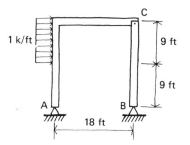

Figure 2-63

2-21. Determine the reaction components for the frame in Figure 2-64.

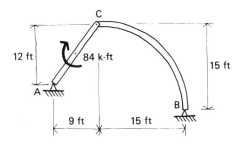

Figure 2-64   Structure acted upon by a couple.

2-22. Determine the reaction components for the structure in Figure 2-65.

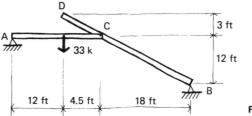

Figure 2-65

2-9 STABILITY AND DETERMINACY

In order to be in a state of static equilibrium, a structure must meet the requirements of stability. Loads and reactions bear no meaningful relationship to one another in an unstable structure. Structural stability is accomplished through the geometry of the members and the support (or boundary) conditions present. First, a stable structure is one that will remain at rest under any realistic loading pattern. For example, the simple beam in Figure 2-66(a) is generally unstable. Even though it might remain at rest under a specific load, such as in Figure 2-66(b), it is still judged unstable. Second, the structure must be capable of carrying load without requiring an angular change in its geometry. The structure held in place by cables in Figure 2-67 is unstable because its load-carrying ability depends upon a change in geometry—in this case, a small motion to the right. The amount of motion is not important. The fact that such motion must take place before the structure can accept load is important.

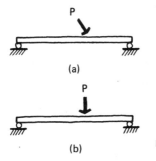

Figure 2-66

It may be helpful to consider the three concurrent force structures in Figure 2-68. The horizontal bars of Figure 2-68(b) are unstable because, as two-force members, they cannot develop the vertical components needed to equilibrate the load P. In other words, an angle change is necessary. But, if we accept the concept of rigid bodies, such an angle change is impossible because the "rigid" bars cannot elongate to accommodate this change.

Structural stability, for the purposes of equilibrium, is a theoretical concept, and it should be remembered that bodies are considered weightless as well as rigid

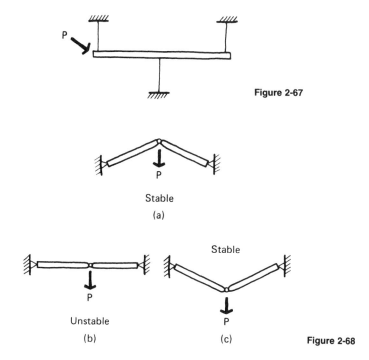

Figure 2-67

Stable

(a)

Stable

Unstable

(b)   (c)   **Figure 2-68**

and that rollers and pins are assumed to be frictionless. Examine each of the structures in Figure 2-69 with respect to the previous discussion. By our rules, the two shown in Figure 2-69(a) and (b) are unstable, while those of Figure 2-69(c) and (d) are stable.

Once a structure has satisfied the conditions of stability, it can then be classified as determinate or indeterminate with respect to its reactions. A member or structure is statically *determinate* if the number of independent reaction components

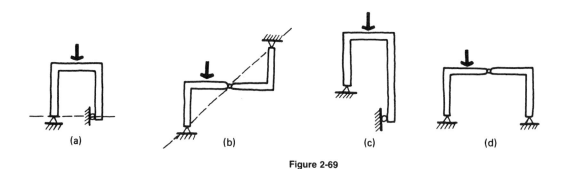

(a)   (b)   (c)   (d)

Figure 2-69

does not exceed the number of applicable independent equations of equilibrium. (This is really just a precise way of stating the familiar axiom, "the number of unknowns cannot exceed the number of equations.") If this is true, then those reaction components can be determined using the techniques of statics alone. If, on the other hand, there exist extra or redundant reaction components, then the structure is said to be statically *indeterminate*. This means that, in order to determine the reactions, the analyst will have to consider more than the basic equilibrium of forces. In general, this means that the deformation or deflection of a structure or member must be examined.

      If an indeterminate structure has only one redundant force component, it is described as being indeterminate to the first degree; with two redundant components, to the second degree; and so on. Some structures, such as a light timber frame of stud and joist construction, have mostly determinate members. Others, such as the typical cast-in-place reinforced concrete frame, are highly indeterminate.

      One basic procedure for analyzing structures with one degree of redundancy is presented briefly in Chapter 9, but a proper treatment of indeterminate structures is beyond the scope of this text. All the examples and problems presented for quantitative solution in this chapter are statically determinate.

PROBLEMS

2-23. Determine whether or not each of the structures in Figure 2-70 is stable. If stable, then ascertain if it is determinate or indeterminate. If indeterminate, to what degree?

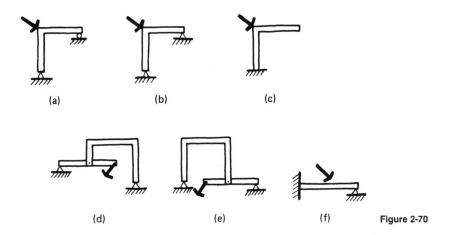

(a)   (b)   (c)

(d)   (e)   (f)   **Figure 2-70**

2-24. Each of the structures in Figure 2-71 is either unstable or indeterminate. Change or remove one of the support conditions in each case so that the resulting structure will be stable and determinate.

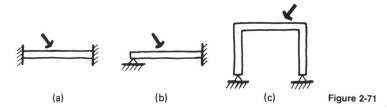

(a)   (b)   (c)   **Figure 2-71**

2-10   SIMPLE CABLE STATICS

A flexible cable will assume a specific geometry when acted upon by one or more point loads. This geometry is dependent upon the relative magnitude and location of each load, the length of the cable, and the height of the supports. The designer usually has little control over the loads but can select the length of the cable, thereby controlling the sag, and the support locations. In general, the greater the sag, the less will be the force in the cable. For reasons of equilibrium discussed previously, a taut cable of very slight sag will have to withstand tremendous internal forces.

Analysis of one-way cable systems is made simple by the fact that a cable has effectively zero moment resistance. It acts like a link chain and, if assumed weightless, will take a straight-line geometry between the loads. Each cable segment then acts like a two-force tension member. Each load point is held in concurrent equilibrium by the load and two internal cable forces, one on each side of the load.

If the loads are applied vertically, the horizontal component of the cable force is a *constant* throughout the cable, and only the vertical component varies from segment to segment. Since the vertical component is dependent upon the cable slope, the largest tension in the cable will occur where the cable slope is greatest, usually at the highest cable support.

Example 2-12

Determine the distance y and the magnitude of the force in each segment of the cable in Figure 2-72.

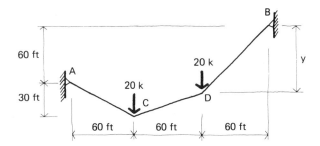

Figure 2-72   Cable with third-point loading.

SOLUTION: First determine the external reactions. The ratio of the components A_x and A_y will be 2:1, following the slope of the cable segment AC.

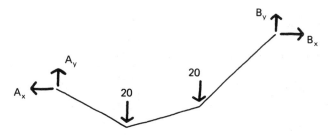

Figure 2-73 Free-body diagram.

Using the FBD in Figure 2-73, we obtain

$$\Sigma M_B = 0$$
$$-A_x(60) - A_y(180) + 20(120) + 20(60) = 0$$
$$-2A_y(60) - 180A_y + 2400 + 1200 = 0$$

$$A_y = 12 \qquad A_y = 12 \text{ kips} \uparrow$$
$$A_x = 24 \qquad A_x = 24 \text{ kips}$$
$$\leftarrow$$

$$\Sigma F_x = 0$$
$$-A_x + B_x = 0$$
$$B_x = 24 \qquad B_x = 24 \text{ kips}$$
$$\rightarrow$$

$$\Sigma M_A = 0$$
$$-20(60) - 20(120) + B_y(180) - 24(60) = 0$$
$$B_y = 28 \qquad B_y = 28 \text{ kips} \uparrow$$

$$\Sigma F_y = 0 \quad \text{(check)}$$
$$12 - 20 - 20 + 28 = 0 \quad \checkmark$$

The distance y must be such that the cable segment BD has a slope that fits the component ratio of B_y to B_x. Therefore,

$$\frac{28}{24} = \frac{y}{60} \qquad y = 70 \text{ ft}$$

The cable force in segment AC is

$$T_{AC} = \sqrt{(12)^2 + (24)^2}$$
$$= 26.8 \qquad\qquad\qquad T_{AC} = 26.8 \text{ kips}$$

The cable force in segment BD is

$$T_{BD} = \sqrt{(24)^2 + (28)^2}$$
$$= 36.9 \qquad\qquad\qquad T_{BD} = 36.9 \text{ kips}$$

The vertical component of the force in CD can be determined by considering its slope or from the vertical equilibrium of either point C or D, as shown in Figure 2-74. The force in segment CD will then be

$$T_{CD} = \sqrt{(24)^2 + (18)^2}$$
$$= 25.3 \qquad\qquad\qquad T_{CD} = 25.3 \text{ kips}$$

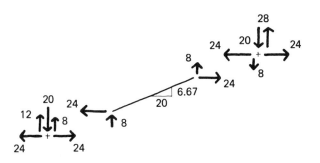

Figure 2-74 Statics of the central portion of the cable in Example 2-12.

When the load points get very close together and are of equal magnitude, the load may be considered to be uniform. If it is uniform along the horizontal projection as in Figure 2-75, the cable will assume a parabolic shape. The maximum sag and its location, labeled y and x, respectively, will depend upon the cable length and the relative support heights as before. Given the value of the uniform load, the designer can vary the span and sag and easily check the influence of such changes on the cable force. With a uniformly loaded cable, the lowest point will usually have a slope of zero, and thus the force in the cable at that point will have no vertical component. This means that, referring to Figure 2-76 and assuming known loads and support locations, we can write two independent moment equations, one about A

and the other about B. The equations will have the same two unknowns, H and x, and can be solved simultaneously. Once x is known, the vertical components of each reaction can be obtained from equilibrium in the vertical direction. Since H is constant throughout the cable, the largest cable force will again be at the uppermost support.

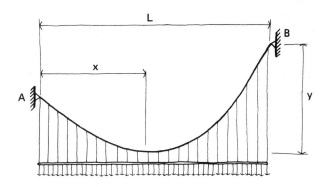

Figure 2-75   Cable with a uniform load.

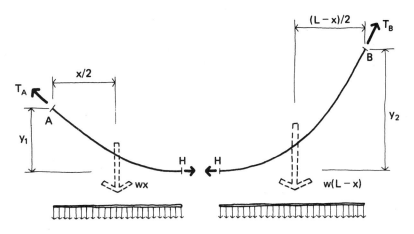

Figure 2-76

Example 2-13

Determine the highest tension in the uniformly loaded cable of Figure 2-77 and Figure 2-78.

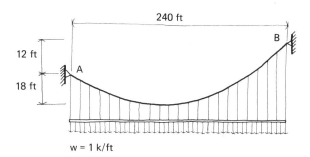

Figure 2-77

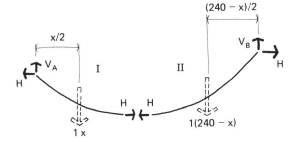

Figure 2-78 Free-body diagrams for Example 2-13.

SOLUTION:

FBD I: $\qquad \Sigma M_A = 0$

$$-1x\left(\frac{x}{2}\right) + H(18) = 0$$

$$18H - \frac{x^2}{2} = 0$$

$$36H - x^2 = 0$$

FBD II: $\qquad \Sigma M_B = 0$

$$-H(30) + 1(240 - x)\left(\frac{240 - x}{2}\right) = 0$$

Solving the two equations will yield the quadratic

$$\frac{2}{5}x^2 + 288x - 34\,560 = 0$$

$$x = \frac{-b \pm \sqrt{b^2 - 4ac}}{2a}$$

$$= \frac{-288 \pm \sqrt{(288)^2 - 4(\frac{2}{5})(-34\ 560)}}{2(\frac{2}{5})}$$

$$= -825 \text{ and } +105$$

The maximum tension will occur at point B. From previous work,

$$H = \frac{x^2}{36}$$

$$= \frac{(105)^2}{36}$$

$$= 306$$

FBD II: $\Sigma F_y = 0$

$$V_B - 1(240 - x) \quad = 0$$
$$V_B - 1(240 - 105) \quad = 0 \qquad V_B = 135$$

$$T_B = \sqrt{(135)^2 + (306)^2}$$

$$= 334 \qquad\qquad\qquad T_B = 334 \text{ kips}$$

Example 2-14

Determine the value of H, the horizontal component of the cable force, in terms of w, L, and y as defined in Figure 2-79. The two cable supports are on the same level.

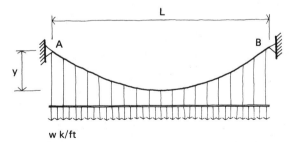

Figure 2-79

w k/ft

SOLUTION: The slope is zero at midspan. Using the free-body diagram in Figure 2-80, we get

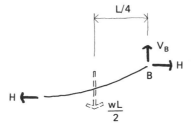

Figure 2-80

$$\Sigma M_B = 0$$

$$\frac{wL}{2}\left(\frac{L}{4}\right) - H(y) = 0$$

$$H = \frac{wL^2}{8y} \tag{2-7}$$

PROBLEMS

2-25. Determine the maximum sag and the largest tension in the symmetrically loaded cable of Figure 2-81.

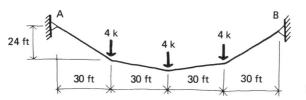

Figure 2-81

2-26. One-half of a suspension structure is shown in Figure 2-82. How much tension will be in the tie-back cable *AB* and how much compression in the mast *BC*?

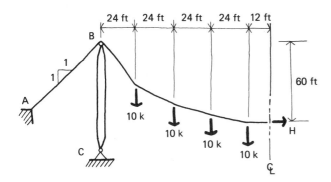

Figure 2-82 Half-elevation.

2-27. A total of 60 kips must be supported by the cable in Figure 2-83. How much should be placed at each of the two load points to achieve the geometry shown?

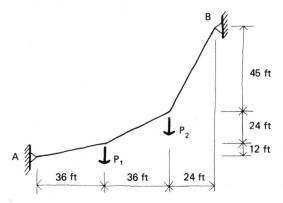

Figure 2-83

2-28. With reference to Figure 2-79, if $L = 300$ ft and $w = 1$ kip/ft, determine the maximum tension in the cable when
(a) $y = 300$ ft
(b) $y = 150$ ft
(c) $y = 75$ ft
(d) $y = 30$ ft
(e) $y = 15$ ft

2-29. Determine the maximum tension in the cable of Figure 2-84.

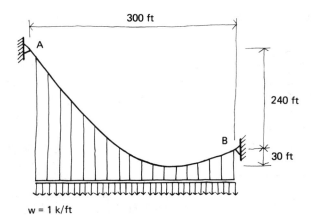

$w = 1$ k/ft

Figure 2-84

2-30. The cables shown in Figure 2-85 carry a uniform load (on the horizontal projection) of 4 kips/ft over the entire 240 ft. Determine the required values of H_L and x in order to maintain equilibrium.

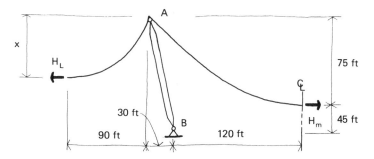

Figure 2-85 Half-elevation of a uniformly loaded cable.

2-11 CONCLUSION AND PROCEDURE

The preceding sections have provided an introduction to concepts of static equilibrium in planar force systems. The principles can easily be extended to three-dimensional systems by increasing the number of equilibrium equations to six, one force equation and one moment equation for each of the three coordinate axes. Problem examples have not been included for three-dimensional systems because no new and fundamental concepts would be involved.

Statics is a subject, like most areas of structural analysis, that cannot be learned by reading about it. The few concepts involved are deceptively simple, but any real understanding comes only through practice. The forces of statics have very real effects on all structures, and the designer must become familiar with the action of such forces and the responses (reactions) made by the structures. While the directions and relative magnitudes of reactive forces are far more important to the architect than their actual quantitative values, no purely qualitative approach to the study of physical forces has proven to be sufficient. It is also important not to confine one's study of the subject to text or classroom examples. The beginning student is urged to attempt to conceptualize or sense the forces that provide equilibrium to the objects in his or her immediate environment. Because of the presence of gravity, such forces can be found everywhere. Begin to examine the statics of ordinary objects such as chairs, doors, signposts, bridges, trees, spider webs, playground equipment, and buildings.

Success in quantitative problem analysis comes only through experience, but that experience should be directed so there is a minimum of wasted effort. The writer's own mistakes and those of his students have resulted in the recommendations that follow.

1.   Always take time to thoroughly review the given situation. Read the problem carefully, noting what it is that you are to find. What is it that the client is asking you to do?

2. To study forces, always sketch free-body diagrams. Do not attempt to work without them. The mere process of sketching can help you to "see" the forces. If you are a visually minded person, you need a picture.

3. Check your assumptions and your answers as you proceed. Make qualitative estimates. Make graphical polygon checks. Are the results rational? Arithmetic may be neat and precise, but common sense is more valuable.

4. Attempt to work slowly and carefully at the beginning of any analysis. Charging into a problem seems to generate more errors than does racing to finish it.

5. Record your work in neat and orderly fashion. Always include units and senses with numerical answers. State assumptions clearly and make notes explaining your procedure as you work. Always work as though someone were going to review your records and follow your steps at a later date. Most often, that someone will be you.

3

structural
properties
of areas

3-1 INTRODUCTION

This chapter is devoted to two concepts that have to do with certain properties of cross sections. In structural analysis, it is necessary to consider more than just the number of square inches included within a cross-sectional area. The shape of the section (or how the material is distributed) is equally important.

In this discussion, the term ''cross section'' is a general one applying to any element or even to a section through an entire structure. It is appropriate to refer to the cross-sectional areas of not only beams and columns, but also trusses, footings, walls, folded plates, segments of shells, and so on.

Two very important concepts are the centroid of an area, which is analogous to the center of mass of a volume; and the moment of inertia of an area, which is most simply described as a measure of resistance to bending and buckling.

3-2 CENTROIDS

The center of mass or center of gravity may be visualized as the location of the resultant of a set of parallel forces. Figure 3-1 illustrates a thin, flat plate of homogeneous material, lying in a horizontal plane. Each element dA, located by coordinates x and y, will be acted upon by a vertical force due to gravity. The resultant of all these parallel forces will be located at the center of gravity of the plate. Conceptually, this point would be the place where we could attach a vertical string to hang the plate and have the plate remain horizontal. It is dimensioned by $\bar{x}$ and $\bar{y}$ in Figure 3-1.

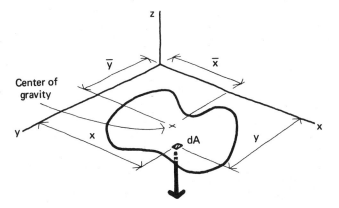

Figure 3-1

If we let the thickness approach zero, the plate becomes an area and the center of gravity is then called a *centroid*. The centroid is like an average location of all the small elemental areas, *dA*. Mathematically, it can be located with respect to any reference axes by two equations.

$$\bar{x} = \frac{\displaystyle\int_0^A x\,dA}{\displaystyle\int_0^A dA} \tag{3-1}$$

$$\bar{y} = \frac{\displaystyle\int_0^A y\,dA}{\displaystyle\int_0^A dA} \tag{3-2}$$

The denominator in each of these expressions is, of course, the total area and the numerator is called the *first moment* of the area, being a summation of areas times distances. The numerator is also sometimes called the *statical moment* of the area.

The concept of the centroid is used in most of the principles found in the study of mechanics of materials and is, therefore, critical to the understanding of structures. As indicated, the center of gravity is an easier concept to grasp, as it can be experimentally determined. For example, if we cut a shape out of cardboard, the shape will balance on the end of a pencil placed under its center of gravity. If the cardboard is uniform in thickness, the center of gravity will be at the centroid of the area. (This experiment will not always be practical because the centroid, and center of gravity for that matter, do not have to be located physically within the area or on the object. For example, the centroid of a doughnut shape would be at the center of the hole.)

For most structural applications, the areas involved are regular geometric shapes, portions of such shapes, or combinations of them. This means that it is seldom necessary to use the calculus to determine a centroid location. Since the centroid of a regular shape is usually known by inspection, the integrals become algebraic sums of the statical moments and areas involved.

$$\bar{x} = \frac{\sum (x_i A_i)}{\sum A_i} \tag{3-1a}$$

$$\bar{y} = \frac{\sum y_i A_i}{\sum A_i} \tag{3-2a}$$

These versions of the basic equations will be used in the examples that follow. Appendix F will be useful in obtaining the centroids of often used shapes.

After the centroid of an area has been determined, it is good practice to sketch a rough scale figure of the area, showing the location of the centroid. Gross errors will appear readily in such a sketch.

Example 3-1

Determine the centroid location of the T shape in Figure 3-2. (*Note*: Unless otherwise indicated, dimensions on cross sections have units of inches throughout this book.)

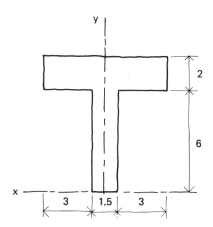

Figure 3-2 T-shaped cross section.

SOLUTION: The shape is comprised of two rectangles with their centroids as shown in Figure 3-3. By symmetry, the centroid of the T will lie along the y axis, and $\bar{x}$ must be zero (Figure 3-4). Note that the statical moment of an area will be zero whenever the reference axis passes through the centroid. The distance, $\bar{y}$, however, has a value between 3 and 7 in. and can be found by Equation (3-2a).

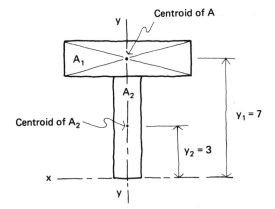

Figure 3-3   Two rectangles in the T shape.

$$\bar{y} = \frac{\sum y_i A_i}{\sum A_i}$$

$$= \frac{y_1 A_1 + y_2 A_2}{A_1 + A_2}$$

$$= \frac{7(2)(7.5) + 3(1.5)(6)}{2(7.5) + 1.5(6)}$$

$$= 5.5 \text{ in.}$$

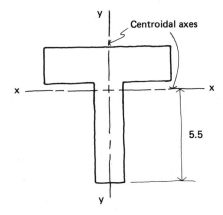

Figure 3-4   Centroid location.

Example 3-2

Find $\bar{x}$ and $\bar{y}$ for the shape in Figure 3-5.

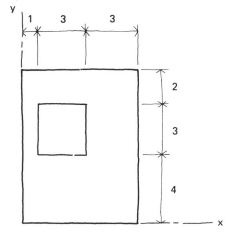

Figure 3-5   Rectangular shape with a square hole.

SOLUTION: Subtract the statical moment and the area of the hole. If we let the subscript 1 represent the area of the rectangle and 2 represent the square hole, we get

$$\bar{x} = \frac{x_1 A_1 - x_2 A_2}{A_1 - A_2}$$
$$= \frac{3.5(7)(9) - 2.5(3)(3)}{7(9) - 3(3)}$$
$$= 3.7 \text{ in.}$$

Similarly, for the y direction,

$$\bar{y} = \frac{y_1 A_1 - y_2 A_2}{A_1 - A_2}$$
$$= \frac{4.5(9)(7) - 5.5(3)(3)}{9(7) - 3(3)}$$
$$= 4.3 \text{ in.}$$

Example 3-3

Determine the $\bar{y}$ for the section shown in Figure 3-7.

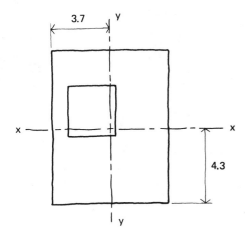

Figure 3-6 Centroid location for Example 3-2.

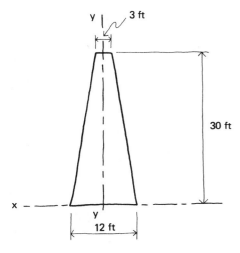

Figure 3-7 Retaining-wall section.

SOLUTION: Assume the area to be made up of two triangles (subscript 2) and one rectangle (subscript 1).

$$\bar{y} = \frac{y_1 A_1 + 2y_2 A_2}{A_1 + 2A_2}$$

$$= \frac{(15)(3)(30) + 2(10)(\frac{1}{2})(4.5)(30)}{3(30) + 2(\frac{1}{2})(4.5)(30)}$$

$$= 12 \text{ ft}$$

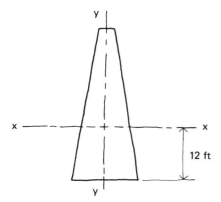

Figure 3-8   Centroid location for Example 3-3.

PROBLEMS

3-1.   Determine $\bar{y}$ for the cross section in Figure 3-9.

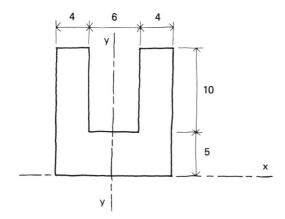

Figure 3-9   Channel beam cross section.

3-2.   Determine the centroid location for the symmetrical angle in Figure 3-10.

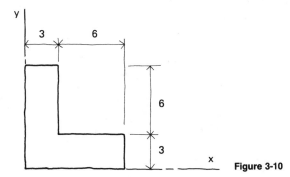

Figure 3-10

3-3. Determine $\bar{y}$ for the shape in Figure 3-11.

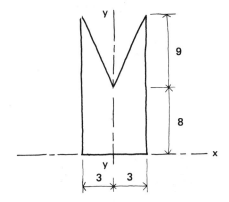

Figure 3-11

3-4. Determine the centroid location for the group of footing pads shown in Figure 3-12. Each square is 9 ft on a side and each circle is 9 ft in diameter.

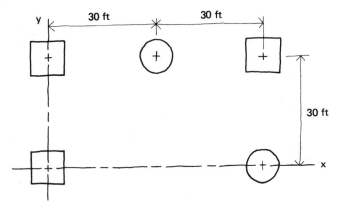

Figure 3-12 Footing plan.

3-5.   Locate the centroid of the prestressed single-T in Figure 3-13.

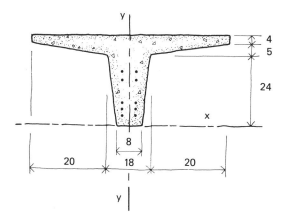

Figure 3-13 Prestressed beam cross section.

3-6.   Figure 3-14 shows a cross section through a timber beam taken where a hole 2 in. in diameter was drilled to permit the passage of a pipe. Determine $\bar{y}$ with respect to the x axis, which was a centroidal axis before the hole was drilled. (*Hint*: First determine $\bar{y}$ with respect to the base of the section.)

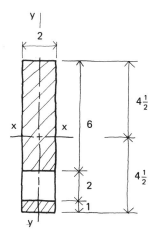

Figure 3-14

3-3 MOMENT OF INERTIA

The *moment of inertia* of an area is a mathematical concept that is used to quantify the resistance of various sections to bending or buckling. It is a shape factor that measures the relative location of material in a cross section in terms of effectiveness. A beam section with a large moment of inertia, or *I* value, will have smaller stresses

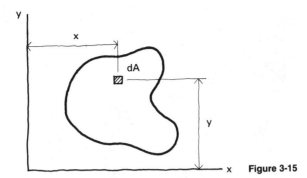

Figure 3-15

and deflections under a given load than one with a lesser I value. A thin shell will have less tendency to buckle if its surface is shaped so that a large moment of inertia is present. A long, slender column will not buckle laterally if the moment of inertia of its cross section is sufficient.

The concept of moment of inertia is vital to understanding the behavior of most structures, and it is unfortunate that it has no accurate physical analogy or description. Mathematically it is easy to compute, and in structural analysis it is easy to use, but beginning students often have difficulty with its abstract nature.

A moment-of-inertia value can be computed for any shape with respect to any reference axis. Figure 3-15 shows this general situation. The moment of inertia of an area about a given axis is defined as the sum of the products of all the elemental areas and the square of their respective distances to that axis. Thus, we get the following two equations from Figure 3-15.

$$I_x = \int_0^A y^2 dA \tag{3-3}$$

$$I_y = \int_0^A x^2 dA \tag{3-4}$$

An I value has units of length to the fourth power, because the distance from the reference axis is squared. For this reason, the moment of inertia is sometimes called the *second moment* of an area. More important, it means that elements that are relatively far away from the axis will contribute substantially more to an I value than those which are close by. Assuming that each element has the same area dA, then one located at twice the distance of another will have four times the moment of inertia.

For structural analysis purposes, usually only two I values are important, the ones that can be computed with respect to the centroidal x and y axes. These are called the *principal axes*. Figure 3-16 shows a simple wood joist of rectangular cross section. Following the previous discussion, this shape has a much larger moment of

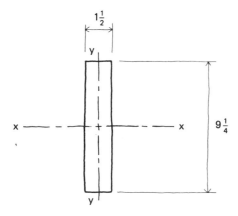

Figure 3-16 A 2 × 10 joist with actual dimensions.

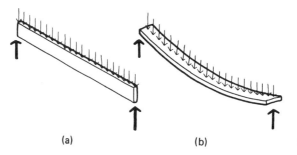

(a)   (b)   Figure 3-17

inertia about the x axis than it does about the y axis. This confirms what we already know from experience, that a rectangular shape has more resistance to bending if used as a joist than if used as a plank. Laid on its side, as in Figure 3-17(b), so that the load would be parallel to the x axis, the rectangle would deflect and break under relatively low load. Like many structural elements, the rectangle has a strong axis and a weak axis. It is far more efficient to load the cross section so that bending occurs about the strong axis. Figure 3-17 illustrates strong-axis bending and weak-axis bending.

It may help to understand the concept of moment of inertia if we draw an analogy based upon real inertia due to motion and mass. Imagine the two shapes in Figure 3-18 to be cut out of heavy sheet material and placed on an axle (xx) so they could spin about it. The two shapes have equal areas, but the one in Figure 3-18(a) has a much higher moment of inertia (I_{xx}) with respect to the axis of spin. It would be much harder to start it spinning, and once moving, much harder to stop. The same principle is involved when a figure skater spins on the ice. With arms held close in, the skater will rotate rapidly, and with arms outstretched (creating increased resistance to spin and/or more inertia), the skater slows down.

Similarly, a beam section shaped as in Figure 3-18(a), with flanges located far from the centroidal axis, will have far more resistance to bending than the cruciform shape in Figure 3-18(b).

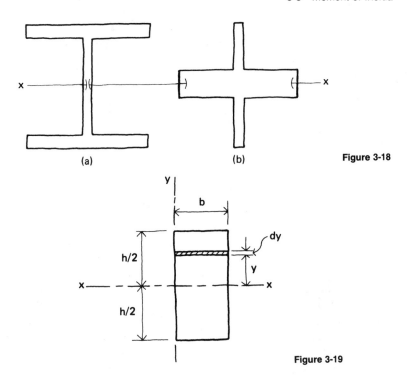

(a)   (b)   **Figure 3-18**

Figure 3-19

As with centroidal determinations, it is seldom necessary to use the calculus to find moments of inertia for the shapes commonly used in building structures. Most often, I values for regular shapes can be expressed in terms of the section dimensions. For example, the centroidal moment of inertia (I_{xx}) of a simple rectangle of dimensions b and h is found from Figure 3-19.

$$I_{xx} = \int_0^A y^2 dA$$

$$= \int_{-h/2}^{h/2} y^2 b \, dy$$

$$= b \left[\frac{y^3}{3} \right]_{-h/2}^{h/2}$$

$$= \frac{b}{3} \left[\frac{h^3}{8} + \frac{h^3}{8} \right]$$

$$I_{xx} \atop \text{centroid} \atop \text{rectangle} = \frac{bh^3}{12}$$

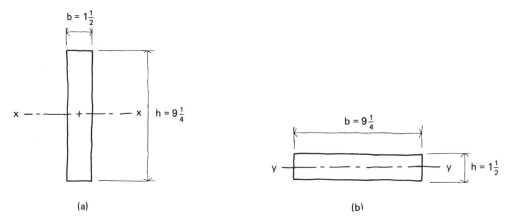

Figure 3-20

The moment of inertia expressions for several geometric shapes appear in Appendix F and will be used in the example problems. The values with respect to an axis located along an edge or base are also given, as these are sometimes useful.

We can now quantify the previous illustration of how a rectangular joist has greater resistance to bending if used upright as opposed to flat. For the joist in question and referring to Figure 3-20, the centroidal moments of inertia are

$$I_{xx} = \frac{bh^3}{12} \qquad\qquad I_{yy} = \frac{bh^3}{12}$$
$$= \frac{1.5(9.25)^3}{12} \qquad = \frac{9.25(1.5)^3}{12}$$
$$= 98.9 \text{ in}^4 \qquad\qquad = 2.60 \text{ in}^4$$

The figures indicate that the nominal 2×10 section has a strong-axis moment of inertia that is almost 40 times larger than its weak-axis moment of inertia. The strong- and weak-axis moments of inertia for this and other timber rectangular sections may be found in Appendix I. The same values for selected steel beam shapes appear in Appendix J.

Example 3-4

Determine the centroidal moments of inertia for the wide-flange shape in Figure 3-21.

SOLUTION: The I_{yy} computation is merely the summation of the I values of three rectangles. To compute I_{xx}, it will be necessary to subtract the I values of two rectangles of "space," located on either side of the web, from the I value of the enclosing rectangle.

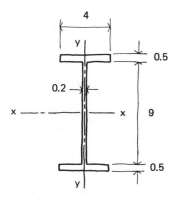

Figure 3-21

$$I_{yy} = [I_{yy}]_{web} + 2[I_{yy}]_{flange}$$

$$= \left[\frac{bh^3}{12}\right]_{web} + 2\left[\frac{bh^3}{12}\right]_{flange}$$

$$= \frac{9(0.2)^3}{12} + 2\left[\frac{0.5(4)^3}{12}\right]$$

$$= 5.34 \text{ in}^4$$

$$I_{xx} = [I_{xx}]_{gross} - 2[I_{xx}]_{space}$$

$$= \left[\frac{bh^3}{12}\right]_{gross} - 2\left[\frac{bh^3}{12}\right]_{space}$$

$$= \frac{4(10)^3}{12} - 2\left[\frac{1.9(9)^3}{12}\right]$$

$$= 102 \text{ in}^4$$

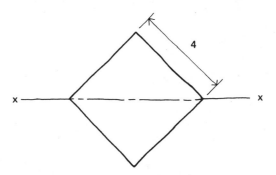

Figure 3-22   Rotated square section.

Example 3-5

Determine the moment of inertia with respect to the *xx* axis of the 4-in. square, which has been rotated 45° as in Figure 3-22.

SOLUTION: The section is comprised of two triangles attached at their bases. From Appendix F, for each triangle,

$$I_{xx_{\text{base triangle}}} \quad \frac{bh^3}{12}$$

The dimensions b and h can be computed because of the 45° relationship.

$$h = 4(\sin 45°)$$
$$= 2.83 \text{ in.}$$
$$b = \frac{4}{\sin 45°}$$
$$= 5.66 \text{ in.}$$
$$I_x = 2\left[\frac{bh^3}{12}\right]$$
$$= 2\left[\frac{5.66(2.83)^3}{12}\right]$$
$$= 21.4 \text{ in}^4$$

Example 3-6

Compute the centroidal moments of inertia of the shape in Figure 3-23.

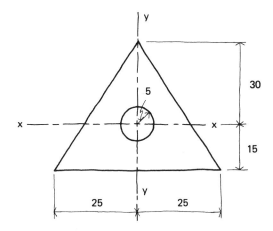

Figure 3-23   Triangle with hole.

SOLUTION: I_{xx} can be obtained by subtracting the I of the circular hole from the I of the gross triangle. (Note that this direct subtraction is only possible because the triangle and the circle share the same xx centroidal axis.) I_{yy} can be computed by subtracting the I of the hole from a larger I computed by adding two values for triangles about a common base. The common base in this case is the vertical yy axis.

$$I_{xx} = [I_{xx}]_{\text{gross}} - [I_{xx}]_{\text{hole}}$$

$$= \left[\frac{bh^3}{36}\right]_{\text{gross}} - \left[\frac{\pi R^4}{4}\right]_{\text{hole}}$$

$$= \frac{50(45)^3}{36} - \frac{\pi(5)^4}{4}$$

$$= 126\ 072\ \text{in}^4$$

$$I_{yy} = 2[I_y]_{\text{base}} - [I_{yy}]_{\text{hole}}$$

$$= 2\left[\frac{45(25)^3}{12}\right] - \frac{\pi(5)^4}{4}$$

$$= 116\ 697\ \text{in}^4$$

PROBLEMS

3-7. Determine the centroidal moments of inertia for the beam cross section of Figure 3-24.

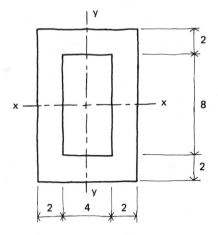

Figure 3-24 Hollow rectangle.

3-8. Determine the centroidal moments of inertia for the T-beam cross section of Example 3-1.

3-9. Determine the I_{xx} value for the tapered section in Figure 3-25.

3-10. Determine the I_{xx} value for the precast plank cross section of Figure 3-26. Each hole is 6 in. in diameter.

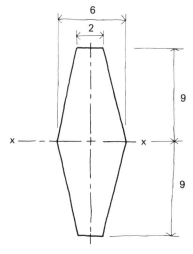

Figure 3-25

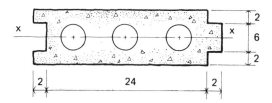

Figure 3-26   Precast plank.

3-11. Determine the moment of inertia, with respect to the centroidal xx axis of the section in Figure 3-27. It is built up from two skins, each ½ in. thick, and six nominal 2 × 6 members. (Actual dimensions are 1½ × 5½ in.)

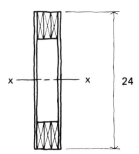

Figure 3-27   Plywood box beam.

3-12. With reference to the T shape of Figure 3-28, first locate the xx centroidal axis by finding $\bar{y}$, and then compute the value of I_{xx}.

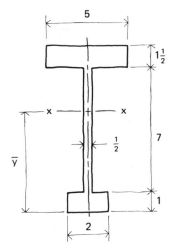

Figure 3-28

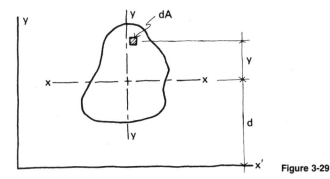

Figure 3-29

3-4 PARALLEL AXIS THEOREM

The addition and subtraction of moment of inertia values for parts of complex shapes can become confusing and lead to errors. The parallel axis theorem provides a simple way to compute the moment of inertia of a shape about any axis parallel to a centroidal one. Its use not only saves time but eliminates most of the confusion. It is easily derived. Assume that we wish to find the moment of inertia of the general shape in Figure 3-29 with respect to an x' axis which is parallel to the centroidal xx one and located d distance away. The general expression will give us

$$I_{x'} = \int_0^A (y + d)^2 dA$$

$$= \int_0^A (y^2 + 2yd + d^2) dA$$

$$I_{x'} = \int_0^A y^2 dA + 2d \int_0^A y\, dA + d^2 \int_0^A dA$$

The integral in the second term of this expression is the statical moment of the shape with respect to one of its own centroidal axes and, as such, must be zero-valued. Since the first term is a centroidal moment of inertia for the shape, the parallel axis theorem is reduced to

$$I_{x'} = I_{xx} + Ad^2 \qquad\qquad (3\text{-}5)$$

where $I_{x'}$ = moment of inertia of the shape about a remote x' axis
$\quad\ I_{xx}$ = moment of inertia of the shape about the centroidal xx axis
$\quad\ A$ = area of the shape
$\quad\ d$ = perpendicular distance between the xx and x' axes

In almost all applications of this theorem, the x' or remote axis is actually the centroidal axis of a composite section made up of several geometric shapes. The parallel axis theorem is applied to each of the shapes in turn to find the total or composite moment of inertia. In every case, the axes of the individual shapes *must* be centroidal ones. The following examples will illustrate the theorem's use.

Example 3-7

Determine the moment of inertia about the xx centroidal axis for the shape shown in Figure 3-30.

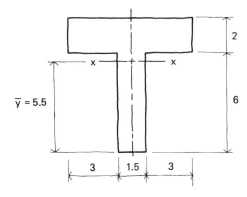

Figure 3-30   T-beam of Example 3-1.

SOLUTION: Since neither rectangle has its centroid coincident with the centroid of the entire section, two applications of the parallel axis theorem will be used. The d distances are as given in Figure 3-31.

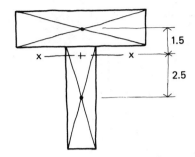

1.5

2.5

x   x

Figure 3-31   Parallel axis distances for Example 3-7.

$$I_{xx} = [I_{xx} + Ad^2]_{\text{flange}} + [I_{xx} + Ad^2]_{\text{stem}}$$

$$= \left[\frac{bh^3}{12} + Ad^2\right]_{\text{flange}} + \left[\frac{bh^3}{12} + Ad^2\right]_{\text{stem}}$$

$$= \left[\frac{7.5(2)^3}{12} + 7.5(2)(1.5)^2\right] + \left[\frac{1.5(6)^3}{12} + 1.5(6)(2.5)^2\right]$$

$$= 122 \text{ in}^4$$

Example 3-8

Determine the I_{yy} value for the retaining-wall section of Figure 3-7.

SOLUTION: The d distances are given in Figure 3-32. The central rectangle

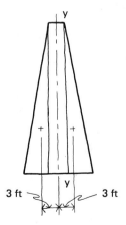

y

y

3 ft   3 ft

Figure 3-32

has its *yy* axis coincident with the *yy* axis of the entire shape, but the triangles do not.

$$I_{yy} = [I_{yy}]_{\text{rectangle}} + 2[I_{yy} + Ad^2]_{\text{triangle}}$$

$$= \left[\frac{bh^3}{12}\right]_{\text{rectangle}} + 2\left[\frac{bh^3}{36} + Ad^2\right]_{\text{triangle}}$$

$$= \left[\frac{30(3)^3}{12}\right] + 2\left[\frac{30(4.5)^3}{36} + \frac{1}{2}(30)(4.5)(3)^2\right]$$

$$= 1434 \text{ ft}^4$$

Example 3-9

To get additional bending resistance, a plate is welded to the top flange of a W14 × 22 as shown in Figure 3-33. The wide-flange shape itself has an I_{xx} value of 199 in^4, an area of 6.49 in^2, and is 13.74 in. in actual depth. Determine the centroidal *xx* moment of inertia for this built-up section.

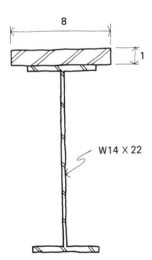

Figure 3-33 Wide-flange beam with plate.

SOLUTION: First determine the location of the centroidal *xx* axis by finding $\bar{y}$. Selecting a reference axis through the bottom edge of the beam, we get

$$\bar{y} = \frac{\Sigma y_i A_i}{\Sigma A_i}$$

$$= \frac{6.87(6.49) + 14.24(8)(1)}{6.49 + 8(1)}$$

$$= 10.9 \text{ in.}$$

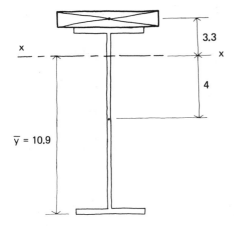

Figure 3-34 Parallel axis distances for Example 3-9.

$I_{xx} = [I_{xx} + Ad^2]_{\text{wide flange}} + [I_{xx} + Ad^2]_{\text{plate}}$. With reference to Figure 3-34, we get

$$I_{xx} = [199 + 6.49(4)^2] + \left[\frac{8(1)^3}{12} + 8(1)(3.3)^2\right]$$
$$= 391 \text{ in}^4$$

PROBLEMS

3-13. Two 2 × 10 joists enclose a 2 × 4 member to make the beam section of Figure 3-35. Determine the centroidal moment of inertia, I_{xx}. (Section properties may be found in Appendix I.)

Figure 3-35 Built-up timber beam.

3-14. Knowing that $\bar{y}$ has a value of 6.34 in. in Figure 3-28, determine I_{xx} for the shape by using the parallel axis theorem.

3-15. Determine the I_{xx} value for the steel box beam of Figure 3-36. It is made of two C10 × 25 and two plates as shown. Each channel has an I_{xx} of 91.2 in^4, an area of 7.35 in^2, and is 10.0 in. in depth.

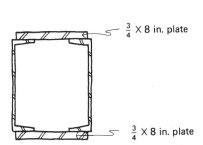

Figure 3-36 Steel box beam.

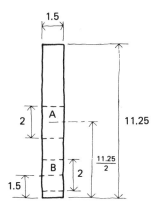

Figure 3-37

3-16. The 2 × 12 joist of Figure 3-37 has an I_{xx} value of 178 in^4, as given in Appendix I. What will be the percentage decrease in this value if a hole 2 in. in diameter is drilled through it at
(a) location A
(b) location B

3-17. The wide-flange beam shape of Example 3-4 must be cut in half at the xx axis to make two structural T shapes, each 5 in. in depth. Compute the I_{xx} value for one of the T sections. What percentage does this represent of the total I_{xx} previously computed in that example?

3-18. Locate the centroidal xx axis and determine I_{xx} for the concrete shape in Figure 3-38.

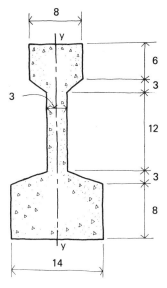

Figure 3-38 Prestressed concrete section.

3-5 RADIUS OF GYRATION

The radius of gyration (r) is a concept that expresses a relationship between the area of a cross section and a centroidal moment of inertia. It is a shape factor, which measures resistance to bending (or buckling) about a certain axis, and accounts for both I and A.

Continuing the concept of rotational inertia discussed in Section 3-3, the radius of gyration represents the location of two parallel lines, one on each side of the axis of spin, at which all of the mass of an object could be concentrated with no change in inertia. For a cross section instead of a mass, we say that all the area may be placed in two lines with no change in the moment of inertia. The r value with respect to a centroidal axis, as indicated in Figure 3-39, is a perpendicular distance from the axis to one of the imaginary lines of concentration. For the shapes most frequently encountered in structural analysis, there are two r values, r_x and r_y, one each for the strong and weak axes, respectively.

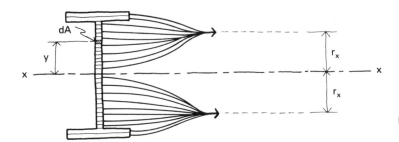

Figure 3-39 Concept of radius of gyration.

If, by definition, the moment of inertia for the shape does not change, then from Figure 3-39,

$$I_{xx} = \int_0^A y^2 dA = r_x^2 \int_0^A dA$$

If we sum up the elemental areas, we get

$$I_{xx} = r_x^2 A$$

Solving for the radius of gyration, we obtain

$$r_x = \sqrt{\frac{I_{xx}}{A}} \qquad (3\text{-}6)$$

Similarly, for the weak axis,

$$r_y = \sqrt{\frac{I_{yy}}{A}} \tag{3-7}$$

The radius of gyration is most useful in the design of slender compression members to resist buckling. It is central to the column buckling theory developed in Chapter 10 but can also be useful in other ways because it has simple units (length) and can replace two section properties.

Example 3-10

Determine the radius of gyration about the xx axis for the T shape of Example 3-7.

SOLUTION:

$$I_{xx} = 122 \text{ in}^4$$

$$r_x = \sqrt{\frac{I_{xx}}{A}}$$

$$= \sqrt{\frac{122 \text{ in}^4}{24 \text{ in}^2}}$$

$$= 2.25 \text{ in.}$$

PROBLEMS

3-19. Determine r_x and r_y for the wide-flange shape of Example 3-4.

3-20. Determine r_x for the rotated square of Example 3-5.

3-21. Figure 3-40 shows two nominal 2 × 6 pieces which are to be fastened together to make a column. What should be the center-to-center spacing s so that r_y will equal r_x for the column cross section?

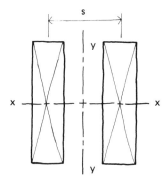

Figure 3-40

4

stress and strain

4-1 TYPES OF STRESS

In simple terms, *stress* is merely the intensity of force. It is force per unit area. The levels and types of stress at different points throughout a building structure are of prime concern to the structural designer. Forces are generated in response to applied loads, dead and live, and those forces always result in a stressed body or an accelerating body. Since most building elements are not meant to accelerate, they develop internal stresses. These stresses, if they become too large, can cause rupture or excess deformation.

Figure 4-1(a) shows a simple bar in tension. The free-body cut in Figure 4-1(b) exposes the stress inside the bar. Even though the loads P are point loads applied along the central axis of the bar, the stress is constant over the cross section except in a zone right next to the end of the bar. The intensity of the stress is simply

$$f_a = \frac{P}{A} \tag{4-1}$$

where f_a = axial stress (psi or ksi)
$\quad\ P$ = axial force (lb or kips)
$\quad\ A$ = cross-sectional area (in^2)

If the stress is too great, it can be reduced by increasing the area of the member. However, it is usually preferable to reduce the force. This can sometimes be done

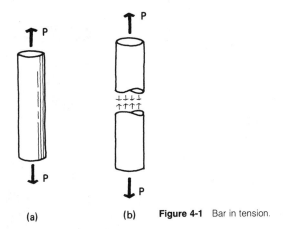

Figure 4-1 Bar in tension.

through a change in geometry or manner in which the loads develop the force (i.e., a change in the statics of the design).

There are really only two kinds of stress, normal and tangential. *Normal stresses* act at right angles to the surface of the stressed area, while *tangential stresses* act parallel to that surface. Shearing stresses are tangential while the axial stress shown in Figure 4-1 is normal. Bending results in both normal and tangential stresses and this will be discussed in detail in Chapters 7 and 8.

PROBLEMS

4-1. Determine the average axial tensile stress in the bar of Figure 4-1 if $P = 50$ kips and the cross-sectional area of the bar is 4 in^2.

4-2. The bolts in Figure 4-2 are being subjected to shear by a vertical reaction. The shear will act tangentially on a transverse section through each bolt. Determine the average shearing stress in the bolts if the reaction is 15 kips and the diameter of each bolt is 1 in.

Figure 4-2

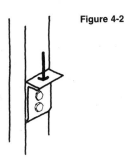

Intuitive understanding of the development of axial stresses and the corresponding strains—and their dependencies upon cross-sectional area and material stiffness—can be used by the architect to promote the reading of a certain kind of character into a building. Such manipulations, though they make use of our visual knowledge of structural behavior, may or may not have anything to do with the actual stability of the structure. Thus the builders of Cologne Cathedral chose to make the front faces of the massive nave piers look like almost-independent columns, the only things visibly connecting the vaults above with the ground below. The "columns" are so slender as to seem incapable of taking any appreciable load from the vaults; they hint, even, at being in tension, not columns at all but cables tying down the upward-ballooning roof. At the other extreme, Frank Furness used visual expectations of normal column proportions, and of behavior under axial compression, to suggest that his façade possesses an almost intimidating weight and density.

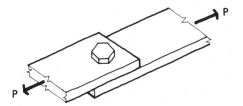

Figure 4-3 Bolted connection.

4-2 BASIC CONNECTION STRESSES

A simple bolted connection can serve to illustrate several different examples of normal and tangential stresses. The bolted connection of Figure 4-3 could fail in any one of four different ways if subjected to overload. Aside from the obvious bolt shear failure shown in Figure 4-4(a), the hole in the plate could elongate by a compression bearing failure as in Figure 4-4(b). In this case the crushing area to be used for design purposes is the projected rectangle, dimensioned by the bolt diameter for one side and the plate thickness for the other. Bearing stress is a normal stress.

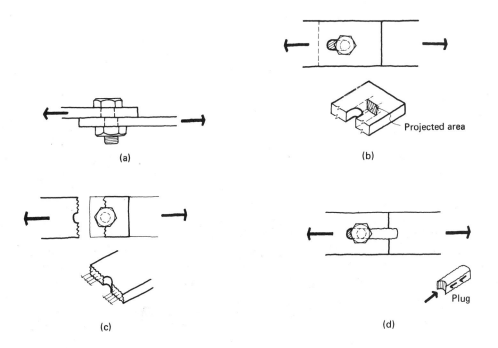

Figure 4-4 Stresses in a bolted connection: (a) shear through the bolt; (b) excess bearing (crushing) of the plate at the hole; (c) tension on the net section of the plate; (d) end shear-out of the plate.

The plate might also fail by excess normal tensile stresses on the area of the plate left after the hole was drilled, as in Figure 4-4(c). This type of stress is called "tension on the net section."

Likewise, if the hole was located too close to the end of one of the plates, the tangential failure of Figure 4-4(d) might result. Here the stressed area is actually two parallel sides of a "plug" of material pushed out of the plate by the bolt.

4-3 STRAIN

Stresses are usually accompanied by a *strain*, which is a physical change in the size or shape of the stressed body. *Normal stresses* result in a shortening or lengthening of the fibers of that body, while *tangential* or *shearing strain* indicates an angle change. It is interesting to note that stress can never be seen, whereas strain can be seen and precisely measured.

Figure 4-5 shows two types of strain. The normal strain in Figure 4-5(a), called δ, is the *total strain*, and in this case it is elongation. This total strain is the sum of the smaller strains occurring in each individual unit length of the bar. The average unit strain is designated as ε.

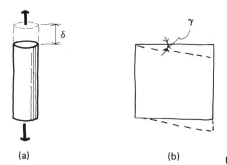

(a)　　　　　　　　(b)　　　**Figure 4-5** Normal strain (a) and shearing strain (b).

Figure 4-6 illustrates a bar with a total strain (shown greatly exaggerated) of 2 in. Each of the 400 units of the original bar got a tiny bit longer, so that the total effect over the aggregate length summed to 2 in. ε is the amount of strain experienced by each unit length. δ is the total strain and equal to ε times the number of units in the original length.

$$\delta = \epsilon L \qquad (4\text{-}2)$$

Solving this for the unit strain, we get

$$\epsilon = \frac{\delta}{L}$$

and for the case in question,

$$\epsilon = \frac{2 \text{ in.}}{400 \text{ in.}} = 0.005$$

Notice that the average unit strain will always be a pure number, as δ and L must have like units.

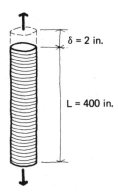

$\delta = 2$ in.

$L = 400$ in.

Figure 4-6

PROBLEMS

4-3. A reinforced concrete column is 15 ft long and under load it shortens ⅛ in. Determine its average unit strain.

4-4. A 300-ft-long steel cable is loaded in tension until the average unit strain is 0.004. Determine the total elongation under this load.

4-4   STRESS VERSUS STRAIN

In 1678, Robert Hooke, an Englishman, observed that most materials were essentially *elastic*, that is, the deformations in a stressed body would disappear upon removal of the load. Furthermore, the relationship between stress and strain (or between load and deformation) was a linear one. Many materials are "springlike," in that if we place a 1-kip load on a member, it will strain a certain amount, and if we add an additional 1-kip load, an additional strain of that same amount will take place. If we continue this loading process, of course, the material will eventually rupture or more usually *yield* (permanently deform), and proportionality between stress and strain will

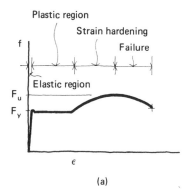

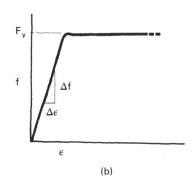

Figure 4-7   Stress-strain curve for mild steel in tension.

no longer exist. Some materials have definite yield points or stress levels beyond which additional load will cause a great increase in strain. Most steels have very definite yield points, and a graph plotting stress versus strain will show a straight line up to the yield stress and then show an abrupt departure from linearity as failure begins. Other materials, such as concrete (in compression), have no definite yield point, and stress is not proportional to strain except at low levels of stress.

Figure 4-7 illustrates the sharp yield point of mild steel. Figure 4-7(b) is merely an enlarged picture of the elastic region in Figure 4-7(a). The ultimate strength of steel is designated as F_u and the yield strength is called F_y. This particular curve is interesting because it shows a large plastic region where strain continues with no increase in load, demonstrating the ''taffylike'' nature of the material. Also indicated is the ''strain-hardening'' property of mild steel, which causes an increase in strength just before failure.

The curve for a structural concrete in compression is shown in Figure 4-8. Notice that the curve for concrete has no real straight-line portion and that stress is only approximately linear with strain and then only for low loads. The ultimate strength of concrete in compression is designated as f'_c.

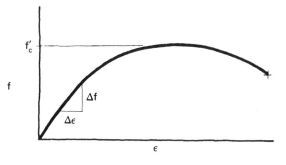

Figure 4-8   Stress-strain curve for concrete in compression.

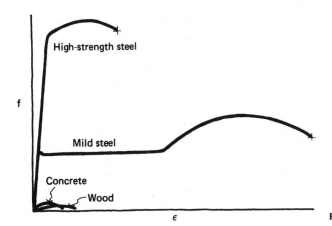

High-strength steel

Mild steel

Concrete

Wood

f

ϵ

Figure 4-9 Stress-strain curves.

Notice that the curves drop downward, indicating a loss of strength near the material failure point. This happens by virtue of the manner in which the materials are tested and merely illustrates the inability of a testing machine to keep load upon a rapidly disintegrating material.

By definition, once a material has experienced any appreciable yield, it will not return to its original shape when the load is removed, but will show some permanent deformation. For all practical purposes, such a material has failed. Therefore, we strive to keep stress levels well below such values when designing structural elements.

The curves of four different materials have been drawn at the same scale in Figure 4-9 so the reader can sense the relative strengths. The curves for concrete and wood are in compression, while those for steel were taken from tensile tests. (Steel tested in compression behaves much as it does in tension, but in practical applications, buckling failures preclude a true compressive yield. Buckling is discussed in Chapter 10.)

4-5 STIFFNESS

The stiffness of a material is a measure of how much that material strains for a given amount of stress. If we place equal loads upon like pieces of wood and steel, the wood will undergo a much larger strain than will the steel. This happens because steel is much stiffer than wood. Because stronger materials are usually stiffer, we tend to confuse stiffness with strength, when in reality the two properties are quite different. The strength of a material can be ascertained as the highest stress achieved during a test of the material (i.e., the largest ordinate at any point on the stress-strain curve). The stiffness, on the other hand, is the slope of the stress-strain curve, given as $\Delta f/\Delta \epsilon$ in Figure 4-7(b) and 4-8. The magnitude of this slope is called the *modulus of elasticity* [or *Young's modulus*, after Thomas Young (1773–1829), an English scientist[1]] and is designated by the symbol E. It has the same units as stress because it is defined as incremental stress divided by incremental strain, and, of course, strain has no units. Values of E for several common materials are tabulated in Appendix E.

The E value or stiffness of concrete is about one-tenth that of steel, while structural wood is about one-half as stiff as concrete. This important property helps to tell us how much a given material will deform or deflect under load. How much will a certain cable stretch? How much will a steel girder deflect? How "bouncy" will a wood joist floor be? The answers to these questions are independent of the material strength but are directly related to material stiffness.

The following example illustrates how E can be quantified.

Example 4-1

A sample of steel is being stressed in a tensile testing machine. Stress is found to be linear with strain in the elastic region and when $f = 7400$ psi, $\epsilon = 0.000\ 25$. At a greater load, $f = 22\ 200$ psi and $\epsilon = 0.000\ 75$. Determine the E value for steel.

SOLUTION:

$$E = \frac{\Delta f}{\Delta \epsilon}$$

$$= \frac{22\ 200 \text{ psi } - 7400 \text{ psi}}{0.000\ 75 - 0.000\ 25}$$

$$= \frac{14\ 800 \text{ psi}}{0.000\ 50}$$

$$= 29.6(10)^6 \text{ psi}$$

(Note: E is a constant for all hot-rolled steel shapes and is usually taken as $29(10)^6$ psi.)

PROBLEMS

4-5. A concrete cylinder with a cross-sectional area of 28.3 in^2 is to be tested in axial compression. Before loading, two marks are scribed on the cylinder precisely 8.000 in. apart, as in Figure 4-10. When the load is 40 000 lb, the marks are measured and found to be 7.997

---

[1]Actually, it was Claude L. M. H. Navier (1785–1836), a French engineer, who first stated the relationship the way we use it today.

in. apart. Determine the *E* value for this concrete. Assume an approximately linear relationship between stress and strain.

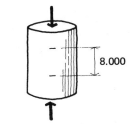

Figure 4-10

4-6. The modulus of elasticity for the steel cable of Problem 4-4 is $25(10)^6$ psi. Determine the average stress in the cable under the applied load. Assume a linear stress-strain curve.

4-6 TOTAL AXIAL DEFORMATION

A unique relationship can be drawn between load and deformation by combining some of the relationships developed in previous sections. The total deformation of an axially loaded member is given by equation

$$\delta = \epsilon L \tag{4-2}$$

The average unit strain ϵ can be expressed in terms of the average unit stress and the modulus, because $E = f/\epsilon$ or $\epsilon = f/E$. Substituting, we get

$$\delta = \frac{fL}{E} \tag{4-3}$$

and since $f = P/A$,

$$\delta = \frac{PL}{AE} \quad \left(\text{HOOKS LAW}\right) \tag{4-4}$$

Equations (4-3) and (4-4) are only valid, of course, if the relationship between stress and strain is essentially linear. For this reason, they are more useful for materials such as steel and less useful for concrete.

Example 4-2

The first-story column of a tall steel building is 20 ft long and must carry an axial load of 5000 kips. If $E = 29(10)^6$ psi and the cross-sectional area of the column is 215 in^2, determine its total shortening.

SOLUTION:

$$\delta = \frac{PL}{AE}$$

$$= \frac{5\ 000\ 000 \text{ lb } (20 \text{ ft})(12 \text{ in./ft})}{29(10)^6 \text{ psi } (215 \text{ in}^2)}$$

$$= 0.192 \text{ in.}$$

or

$$\delta \approx 0.2 \text{ in.}$$

PROBLEMS

4-7. A 500-ft-long roof cable cannot be permitted to stretch more than 30 in. or the roof geometry will change too greatly. If $E = 26(10)^6$ psi and the force in the cable is 1700 kips, determine the required cable diameter needed to avoid excessive elongation.

4-8. A 25-ft-long wood column is stressed in compression to 700 psi. If $E = 1.4(10)^6$ psi, determine the total shortening of the column.

4-7   THERMAL STRESSES AND STRAINS

Whenever a structure is heated or cooled, it changes shape. Most materials expand when heated and contract when cooled. Long building structures must have joints in them to allow these changes to take place. Exposed exterior structural elements often undergo great temperature variations compared to interior ones and large differential movements can result. If these movements are restrained, stresses can build up in the members themselves and connecting elements.

In many building structures, the small range of ambient temperatures and/ or the absence of long uninterrupted structural elements serve to minimize thermal effects and they can be safely ignored. It is helpful to study the magnitudes of thermal movements and stresses to be able to judge when they deserve design attention.

The total change in length of a body due to a temperature change is

$$\delta = \alpha \Delta t (L) \tag{4-5}$$

where δ = total change in length (in.)
  α = thermal coefficient for the material (°F^{-1})
  Δt = temperature change (°F)
  L = original length (in.)

The coefficient α is usually expressed in terms of strain per degree of temperature change (e.g., in./in. per degree Fahrenheit). With Δt expressed in degrees Fahrenheit, it then becomes convenient to think of $\alpha \Delta t$ as equivalent to ϵ, the average unit strain.

Example 4-3

A 125-ft-long masonry wall undergoes a temperature rise from 32°F to 120°F. Determine the total change in length if α for the material is $3.4(10)^{-6}$ in./in. per °F.

SOLUTION:

$$\delta = \alpha \,\Delta t(L)$$

$$= [3.4(10)^{-6} \,°F^{-1}](88°F)(125 \text{ ft})(12 \text{ in./ft})$$

$$= 0.45 \text{ in.}$$

Without intermediate joints to allow movement in the plane of the wall, this motion will accumulate, possibly causing damage to attached walls or other constructions. The real difficulties, however, will develop during the contraction or cooling cycle of the same wall. Masonry and concrete lack the requisite tensile strength needed for large thermal contractions, and vertical cracks will develop in a long unjointed wall.

Example 4-4

A high-rise building 700 ft tall has an exposed steel frame. On a sunny day in winter, the columns on the south side reach 120°F while those on the north side remain at 10°F. Under these extreme conditions, what will be the overall difference in length of the columns? Assume that $\alpha = 6.5(10)^{-6} \,°F^{-1}$.

SOLUTION:

$$\delta = \alpha \,\Delta t(L)$$

$$= [6.5(10)^{-6} \,°F^{-1}](110°F)(700 \text{ ft})(12 \text{ in./ft})$$

$$= 6.0 \text{ in.}$$

Unlike the contraction of a material such as masonry, unrestrained thermal elongation seldom results in a structural failure of the piece itself. However, as soon as that element is attached to others or restrained in any way, stresses will develop. A fully restrained bar fixed at the ends will build up compressive stress while attempting to elongate during a temperature increase. The resulting stress will have the same magnitude as if the bar had been allowed to expand and then axially loaded until it was "squeezed" back to its initial length. In other words, the change in length under free thermal expansion would be

$$\delta_{\text{case 1}} = \alpha \,\Delta t(L)$$

The change in length due to an applied compressive load will be

$$\delta_{\text{case } 2} = \frac{PL}{AE}$$

or

$$\delta_{\text{case } 2} = \frac{fL}{E}$$

Since the actual elongation for a "fixed" end bar is zero,

$$\delta_{\text{case } 1} - \delta_{\text{case } 2} = 0$$

or

$$\delta_{\text{case } 1} = \delta_{\text{case } 2}$$

$$\alpha \, \Delta t(L) = \frac{fL}{E}$$

Solving for the axial stress, we get

$$f = E\alpha \, \Delta t \qquad \qquad (4\text{-}6)$$

This stress would, of course, be tensile if we cooled the bar instead of heating it. Equation (4-6) is interesting in that it illustrates that axial stresses (assuming a constant cross section), which develop in a restrained body due to a temperature change, are independent of any dimension of the body.

Example 4-5

A straight concrete bridge is restrained by two canyon walls. If $E = 3.5(10)^6$ psi and $\alpha = 5.5(10)^{-6}$ °F^{-1}, determine the compressive stress developed during a temperature increase of 60°F.

SOLUTION:

$$f = E\alpha \, \Delta t$$

$$= 3.5(10)^6 \text{ psi } [5.5(10)^{-6} \text{ °F}^{-1}](60\text{°F})$$

$$= 1155 \text{ psi}$$

PROBLEMS

4-9. An unloaded steel roof cable is 400 ft long at 75°F. Determine its length at 10°F and at 140°F. Assume that $\alpha = 6.0(10)^{-6}$ °F^{-1}.

4-10. A long concrete bearing wall has vertical expansion joints placed every 70 ft. Determine the required width of the gap if it is wide open at 20°F and just barely closed at 110°F. Assume that $\alpha = 5.5(10)^{-6}$ °F^{-1}.

4-11. A large steam pipe is built with no provision for thermal expansion. If the ends are fixed, what is the level of compressive stress developed as the pipe goes from 60°F to 240°F? Assume that buckling does not occur. Assume that $E = 29(10)^{6}$ psi and $\alpha = 6.5(10)^{-6}$ °F^{-1}.

4-12. An expansion loop is placed in the pipe of Problem 4-11 as shown in Figure 4-11. The loop is 6 ft long at 60°F. How long is it at 240°F?

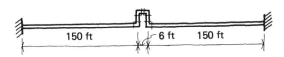

150 ft   6 ft   150 ft

Figure 4-11   Pipe with expansion loop.

5

properties of structural materials

5-1 INTRODUCTION

It is important for the structural designer to realize that different engineering materials have different characteristics and will exhibit different behaviors under load. A knowledge of such characteristics or properties will help to ensure proper use of these materials, both architecturally and structurally.

It is assumed that the reader will have already been exposed to the study of materials through courses in building construction or materials science. This chapter will only highlight a few selected structural materials in the interest of emphasizing the range of structural characteristics and their diversity. A table of properties of selected structural materials is given in Appendix E.

5-2 NATURE OF WOOD

Wood is a natural material and has a broad range of physical properties because of the different characteristics of its many species. Softwoods such as fir, pine, and hemlock are most often used for structural applications because they are more plentiful (grow fast and tall) and are easier to fabricate. These woods are generally strong in tension and compression in a direction parallel to the grain and weak when stressed perpendicularly to the grain. Wood is also weak in shear because of its tendency to split along the natural grain laminations. The allowable stresses for a few selected species are given in Appendix H.

114

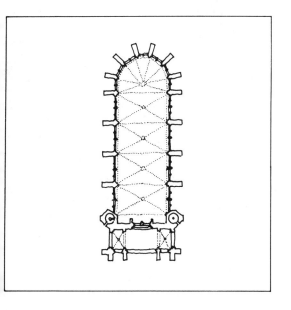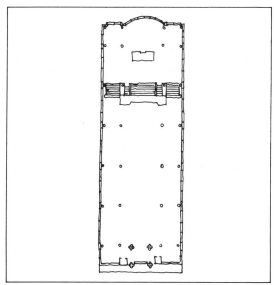

In considering the properties of structural materials, it is important for the architect to realize that although an understanding is necessary to make a successful building, the materials themselves do not usually *dictate* architectural form. This may be illustrated by two contrasting sets of comparisons. Here are plans and interior views of four churches with generally the same architectural intent; the Ste. Chapelle in Paris, constructed in stone; Notre-Dame du Raincy, outside Paris, built in concrete by Auguste Perret (on this page); and Otto Bartning's Steel Church in Germany; and Richard Munday's wooden Trinity Church in Newport, Rhode Island (on the following page).

115

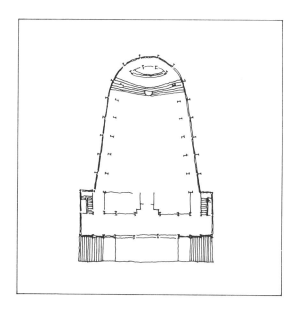

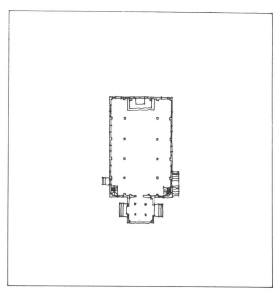

Disregarding the variation in size of the four buildings, they are very similar in plan. In fact on the basis of plan alone, it would be hard to tell which church is of which material. The reason is this; despite the fact that all four seek to make the lightest possible impression on the interior, none of the designs strains the structural capacities of its material. Formal desires, rather than the structural properties of materials, by and large determine the nature of architectural mass and space relations.

    In bridges, on the other hand, conditions are generally more extreme, formal preconceptions fewer, and concern for economy of materials high, and so bridge forms and the characteristics of the materials making the forms are often highly correlated. Stone works best in compression and is conveniently transported in pieces, so arched bridges of stone are a natural result. Wooden bridges take into account the facts that their material will come in sticks, of greater or lesser, but certainly of limited, size, and will take compression or tension with equal capacity. Suspension bridges are unthinkable without some homogeneous material, quite good in tension and available in extreme lengths, like steel cable or rope. Spans in reinforced concrete, at least the best of them, take advantage of its compressive strength and its ability to make large continuous elements and smooth connections.

Wood is light and soft compared to most other structural materials and is easily shaped and fastened together. A minimum of materials-handling equipment is needed to erect wood structures because of their weight. It is also very versatile in terms of its adaptability to the making of geometric shapes and even nonlinear forms.

Most softwoods are fairly ductile and will not fail suddenly when overloaded. Because of their lack of homogeneity or uniformity, the allowable stresses are quite low compared to failure stresses. Consequently, when wood structures are properly engineered, a statistically high margin of safety is present. Wood is often known as the ''forgiving'' material because of its apparent ability to sustain loads not accounted for when the structure was designated.

Wood, on the other hand, is not very stiff. It is subject to excessive deflection and creep deformation if not designed with these characteristics in mind. It is prone to damage by fire and to deterioration by moisture and insects. It expands and contracts with variations in humidity, markedly so in the direction perpendicular to the grain. Timber structures that are to be exposed to the elements must be carefully treated or highly maintained to preserve their integrity.

The American Institute of Timber Construction publishes the *Manual of Timber Construction* and the reader is referred to it for more extensive information on the properties and use of wood. The National Forest Products Association publishes some excellent design aids and data books for use by the structural designer.

5-3 CONCRETE AND REINFORCED CONCRETE

Concrete is a man-made conglomerate stone composed of essentially four ingredients: portland cement, water, sand, and coarse aggregate. The cement and water combine to make a paste that binds the sand and stones together. Ideally, the aggregates are graded so that the volume of paste is at a minimum, merely surrounding every piece with a thin layer. Most structural concrete is stone concrete, but structural lightweight concrete (roughly two-thirds the density of stone concrete) is becoming increasingly popular.

Concrete is essentially a compressive material having almost no tensile strength. As explained in Chapter 8, shearing stresses are always accompanied by tension, so concrete's weakness in tension also causes it to be weak in shear. These deficiencies are overcome by using steel bars for reinforcement at the places where tensile and shearing stresses are generated. Under load, reinforced concrete beams actually have numerous minute cracks which run at right angles to the direction of major tensile stresses. The tensile forces at such locations are being taken completely by the steel ''re-bars.''

The compressive strength of a given concrete is a function of the quality and proportions of its constituents and the manner in which the fresh concrete is cured. (*Curing* is the process of hardening during which time the concrete must be prevented from "drying out," as the presence of water is necessary for the chemical action to progress.) Coarse aggregate that is hard and well graded is particularly essential for quality concrete. The most important factor governing the strength, however, is the percentage of water used in the mix. A minimum amount of water is needed for proper hydration of the cement. Additional water is needed for handling and placing the concrete, but excess amounts cause the strength to drop markedly.

These and other topics are fully covered in the booklet, "Design and Control of Concrete Mixtures," published by the Portland Cement Association. This is an excellent reference, treating both concrete mix design and proper construction practices. The American Concrete Institute publishes a widely adopted code specifying the structural requirements for reinforced concrete.

Concrete is known as the "formable" or "moldable" structural material. Compared to other materials, it is easy to make curvilinear members and surfaces with concrete. It has no inherent texture but adopts the texture of the forming material, so it can range widely in surface appearance. It is relatively inexpensive to make, both in terms of raw materials and labor, and the basic ingredients of portland cement are available the world over. (It should be noted, however, that the necessary reinforcing bars for concrete may not be readily available in less-developed countries.)

The best structural use of reinforced concrete, in terms of the characteristics of the material, is in those structures requiring continuity and/or rigidity. It has a monolithic quality which automatically makes fixed or continuous connections. These moment-resistant joints are such that many low-rise concrete buildings do not require a secondary bracing system for lateral loads. In essence, a concrete beam joins a concrete column very differently from the way steel and wood pieces join, and the sensitive designer will not ignore this difference. (These remarks do not apply to precast structural elements, which are usually not joined in a continuous manner.)

Concrete is naturally fireproof and needs no separate protection system. Because of its mass, it can also serve as an effective barrier to sound transmission.

In viewing the negative aspects, concrete is unfortunately quite heavy and it is often noted that a concrete structure expends a large portion of its capacity merely carrying itself. Attempts to make concrete less dense, while maintaining high quality levels, have generally resulted in increased costs. Nevertheless, use of lightweight concrete can sometimes result in overall economies.

Concrete requires more quality control than most other building materials. Modern transit-mixed concrete suppliers are available to all U.S. urban areas and the mix is usually of a uniformly high quality. Field- or job-mixed concrete requires knowledgeable supervision, however. In any type of concrete work, missing or mislocated reinforcing bars can result in elements with reduced load capacities. Poor handling and/or curing conditions can seriously weaken any concrete. For these and other reasons, most building codes require independent field inspections at various stages of construction.

Proper concrete placement is also somewhat dependent upon the ambient weather conditions. Extremely high temperatures and, more important, those below (or near) freezing can make concrete work very difficult.

5-4   STRUCTURAL STEEL

Steel is the strongest and stiffest building material in common use today. Relative to wood and concrete, it is a high-technology material made by highly refined and controlled processes. Structural steel has a uniformly high strength in tension and compression and is also very good in shear. It comes in a range of yield strengths made by adjusting the chemistry of the material in its molten state. It is the most consistent of all structural materials and is, for all practical purposes, homogeneous and *isotropic*, meaning it has like characteristics in all directions. (By contrast, wood is *anisotropic*.)

The greatest asset to steel is its strength and "plastic reserve," as shown in Figure 4-7. It is highly ductile and deforms greatly before failing if overloaded. Because of steel's strength, the individual members of a frame are usually small in cross section and have very little visual mass. Steel is a linear material and can be economically made into a visual curve only by using a segmented geometry. It is most appropriately used in rectilinear structures where bolted or welded connections are easy to make. The structural shapes (i.e., pipes, tubes, channels, angles, and wide-flange sections) are manufactured to uniform dimensions having low tolerances. They are fully prepared (cut, trimmed or milled, drilled or punched, etc.) in a fabrication shop, remote from the site, and then delivered ready for erection. Such structures go up rapidly with a minimum of on-site labor. The most popular form of construction used today is referred to as shop-welded, field-bolted. In this method the various clip angles, beam seats, and so on, are welded to the members in a shop and then the members are bolted together in the field.

A major disadvantage to structural steel is its need to be fire-protected in most applications. It loses its strength at around 1100°F and will then yield rapidly under low loads. A few municipalities require that all structural steel be fire-protected, and most codes will not permit any exposed elements to be within approximately 12 ft of a combustible fire source.

The making of steel requires large physical plants and a high capital outlay, and therefore relatively few countries of the world have extensive mill facilities. The cost of manufacturing, coupled with the cost of transportation, can make steel a relatively expensive material. Just the same, in most urban areas, concrete and steel are quite competitive with one another in terms of in-place construction costs.

Continuity in the connections is much harder to achieve in steel than in concrete, and most buildings are constructed with simple connections or ones that are only partially moment-resistant. Some type of lateral load bracing system is almost always required in a steel-framed building and must be considered early in the design process.

Rolled steel is manufactured in a wide range of strengths. The standard low-carbon mild steel in use today has a yield strength of 36 ksi. Steel plate can be obtained with an F_y value of 100 ksi, and most standard shapes can be rolled in steel as strong as 65 ksi, although this can be expensive. Examples and problems in this text are limited to shapes of $F_y = 36$ ksi and $F_y = 50$ ksi. These particular values are the most common ones, with the lower 36-ksi strength being the most frequently specified.

Information about the various kinds of steel available can be obtained directly from manufacturers and fabricators. The reader is also advised to purchase the latest edition of the *Manual of Steel Construction*, published by the American Institute of Steel Construction. It is an indispensable reference work for the design professional.

5-5 MASONRY AND REINFORCED MASONRY

Like concrete, brick and concrete masonry units are strong in compression and weak in tension. These materials have traditionally been used in walls, both bearing and nonbearing. Usually, wall thicknesses required by code specifications to prevent lateral instability are such that the actual compressive stresses are low. Crushing is seldom an important design constraint.

Masonry walls are more permanent than wood walls and provide effective barriers to both fire and noise. They are less expensive and often more attractive than formed concrete walls. Brick generally has more variation of pattern and texture than does concrete block, but is also more expensive.

It is becoming increasingly common to use reinforced concrete block for retaining walls and structural pilasters. In this construction, individual reinforcing bars are grouted in some of the vertically aligned cells of the concrete units and serve as tensile reinforcement. This greatly increases the lateral load capacity of the block. Reinforcing can also be placed in special channel-shaped blocks to serve as lintels and tie beams. Brick can be reinforced by using two wythes to create a cavity for grout and reinforcing bars. The brick not only serves as formwork but also carries compressive forces under load.

5-6 CREEP

Section 4-3 explained how structural elements change their size and shape upon application of load. This is called *elastic strain* and, provided that we do not stress the material too greatly, such deformation will disappear upon removal of the load. Most materials, if left under load for a long time, will exhibit an additional strain referred to as *creep*. In some cases these strains will remain after removal of the load.

The amount of creep, which takes place under long-term load, seems to vary directly with the stress level present and the ambient temperature and inversely with the material stiffness. Many plastics creep considerably in just a short period of time. Steel exhibits very little creep except at elevated temperatures. Concrete and wood both creep appreciably if stressed highly for long periods of time.

Members that must support constantly applied loads such as dead weight should be "overdesigned" so that the stresses will be low. For example, the increased deflection (over a couple of years) of a reinforced concrete beam carrying a heavy masonry wall can be double the initial elastic deflection. Many cantilevered portions of wood structures develop an unsightly sag with time which could have been prevented or minimized through the proper consideration of creep.

6

shear
and
moment

6-1 DEFINITIONS AND SIGN CONVENTIONS

A transverse load on a linear element such as a beam or column will generate essentially two kinds of stress, shearing and flexural. *Shearing stresses* are the result of internal shearing forces, and *flexural stresses* result from internal resisting couples or moments. Both of these effects are responses to the externally applied forces and will vary along the length of a member. Their magnitudes and senses at any section will be dependent upon the loads, span, and support conditions of the member.

Usually, the structural designer is interested in the maximum values of these shears and moments, which exist in the many elements or parts of a structure. These maximum values will help to make judgments as to the soundness of the overall scheme and in the planning of the geometry of the various structural elements. Member sizes are also determined and spans sometimes modified on the basis of maximum shear and moment values.

In order to study these forces, we use free-body diagrams and the techniques of statics. By cutting a transverse section through a beam or column, we expose the internal forces and make them external via a free-body diagram (see Figure 2-34). Statics can then be applied to solve for the magnitudes and senses of the unknown values. Graphs or plots are made to show how forces change from section to section. A plot illustrating how the internal shearing force V varies over the length of a member is called a *shear diagram* or *V diagram*. A *moment diagram* or *M diagram* is a similar plot showing the variation of the internal moment throughout the member. Shears and moments are plotted according to the following sign conventions.

Figure 6-1 Sign convention for shearing forces.

Shearing forces that tend to cause the slippage failures shown in Figure 6-1 will be denoted by the sign accompanying them. Notice in Figure 6-2 that for positive shear, forces exist (for equilibrium) which are *up* on the left-hand face and *down* on the right-hand face of an element. The opposite forces act on the cut faces when they are subjected to negative shear. (*Note*: The sign convention for shear is somewhat arbitrary and the opposite plus and minus associations are preferred by some writers.)

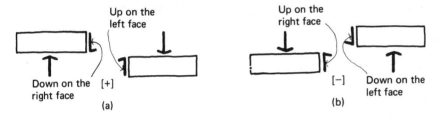

Figure 6-2 Internal shearing forces.

The sign convention for internal bending moment is considerably more straightforward, as Figure 6-3 shows. Positive moment generates concave upward curvature, causing compression in the top fibers and tension in the bottom fibers. Negative moment causes concave downward curvature and, of course, the opposite types of fiber strain. This convention is the standard one for curvature in mathematics and is universally accepted. Since the convention is related to strain, it is possible to look at the probable deflected shape of a beam under load, for example, and

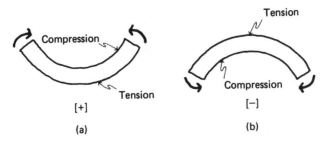

Figure 6-3 Sign convention for internal moment forces.

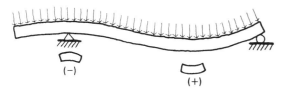

Figure 6-4   Beam curvature.

determine what portions of the span have positive or negative internal moments. The uniformly loaded beam in Figure 6-4 has an overhang, and from the deflected shape we can see that negative internal moments exist over part of the beam's length and positive internal moments are present in another portion. The implication here is that there is likely a transverse section or portion of the span where the bending moment is zero to accommodate the required sign change. Such a section, termed a *point of inflection*, is almost always present in overhanging beams.

The most important feature of these sign conventions is that they are different from the conventions used in statics. When using the three equations of equilibrium, forces up and to the right are plus and counterclockwise moments are plus. The new sign conventions are used only for plotting the shear and moment diagrams. It is important that the two conventions not become confused.

6-2   SHEAR AND MOMENT EQUATIONS

The most basic way to obtain V and M diagrams is to graph specific values from statics equations which have been written so they are valid for appropriate portions of the member. (In these explanations we shall assume that the member is a beam, acted upon by downward loads, but actually the member could be turned at any angle. In the general case, we are determining shears and moments due to transverse loads.) The following examples will illustrate how we can write and plot V and M equations. In each case, the uniform load of the beam's own weight has been neglected.

Example 6-1

Construct the V and M diagrams for the beam in Figure 6-5.

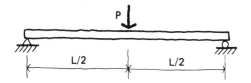

Figure 6-5   Simple beam with a concentrated load at midspan.

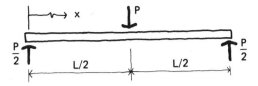

Figure 6-6

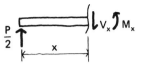

Figure 6-7

SOLUTION: If we let x designate any point along the length of a beam and assume the origin to be at the left end ($x = 0$ there), we can write expressions for V and M in terms of constants (the external loads and reactions) and the distance x. To examine V_x and M_x in the left-hand half of the beam in Figure 6-6, we cut through that portion. This will make the unknowns external and we can use statics on the resulting free body of Figure 6-7. Notice that the unknowns have been assumed positive by the new convention. This will mean that answers yielded by statics which come out plus (as assumed) will be plotted as positive ordinates, and those that turn out with minus signs (not as assumed) must be plotted as negative values. Now, applying the equations of equilibrium to find the unknowns, we get

$$\Sigma F_y = 0$$
$$\frac{P}{2} - V_x = 0$$
$$\left(0 \leqslant x \leqslant \frac{L}{2}\right) \quad V_x = \frac{P}{2}$$

Taking moments at the cut face (thereby eliminating V_x), we get

$$\Sigma M_x = 0$$
$$-\frac{P}{2}(x) + M_x = 0$$
$$\left(0 \leqslant x \leqslant \frac{L}{2}\right) \quad M_x = \frac{P}{2}x$$

For the right-hand half and using the free body in Figure 6-8,

Figure 6-8

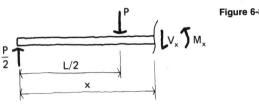

$$\Sigma F_y = 0$$

$$\frac{P}{2} - P - V_x = 0$$

$$\left(\frac{L}{2} \leqslant x \leqslant L\right) \qquad V_x = \frac{P}{2} - P$$

$$\Sigma M_x = 0$$

$$-\frac{P}{2}(x) + P\left(x - \frac{L}{2}\right) + M_x = 0$$

$$\left(\frac{L}{2} \leqslant x \leqslant L\right) \qquad M_x = \frac{P}{2}(x) - P\left(x - \frac{L}{2}\right)$$

Substituting finite values of x (e.g., $x = 0$, $L/4$, $L/2$, $3L/4$, and L), we can plot the equations as shown in Figure 6-9. In this case the V diagram does not vary with x except in sign. The ordinates on the moment diagram are all positive, as might be verified by the deflected shape of this beam.

From the diagrams we can also see the necessity for writing an equation for each half of this beam. Any type of load change or application (including reactions)

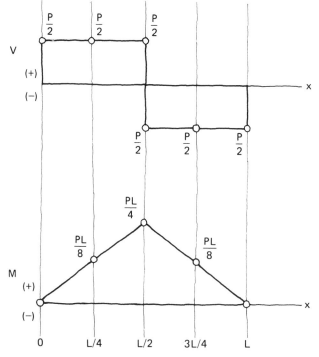

Figure 6-9   *V* and *M* diagrams for Example 6-1.

will cause discontinuities in the diagrams so that each set of V and M equations is only valid for a specific part of the beam length. However, equations with appropriate interval limits can always be written for each beam portion.

Example 6-2

Construct the V and M diagrams for the beam in Figure 6-10.

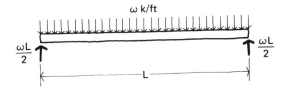

Figure 6-10 Simple beam with a uniform load.

SOLUTION: Since the uniform load is constant over the entire span, only one set of equations will be necessary. Using Figure 6-11, we get

$$\Sigma F_y = 0$$

$$\frac{wL}{2} - wx - V_x = 0$$

$$(0 \leq x \leq L) \qquad V_x = \frac{wL}{2} - wx$$

$$\Sigma M_x = 0$$

$$\frac{-wL}{2}(x) + wx\left(\frac{x}{2}\right) + M_x = 0$$

$$(0 \leq x \leq L) \qquad M_x = \frac{wLx}{2} - \frac{wx^2}{2}$$

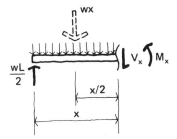

Figure 6-11 Free-body diagram.

Substituting the values of x equal to the quarter points of the span, we get the diagrams shown in Figure 6-12. Ordinates that lie above the reference line are taken as positive and those below, as negative.

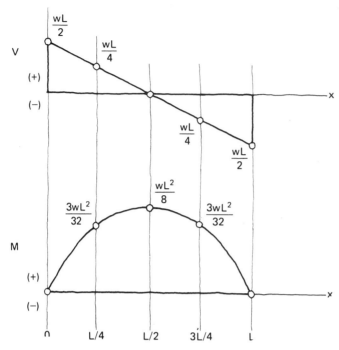

Figure 6-12 *V* and *M* diagrams for Example 6-2.

The shear diagram reflects the linear variation with *x* as required by the shear equation. The moment diagram is a parabolic curve, which follows the second-power function of *x* present in the moment equation. This can be confusing, as one is led to think that bending is a second-power function of the span alone. Moment is always linear with span; it is just that with a uniform load, as opposed to a concentrated one, a change in length means a change in load. For this reason the plot becomes parabolic. To understand this better, compare the maximum moment for the concentrated load of Example 6-1 (i.e., $M = PL/4$) to the maximum moment for the uniform load, which is $M = wL^2/8$. Since *P* is the total load in the first case, let $wL = W$ to get a comparable load for the uniform case. Then *M* will be $WL/8$, indicating that moment varies linearly with span and linearly with load. This also shows clearly that concentrated loads will generate double the moment caused by uniform loads.

Example 6-3

Construct the *V* and *M* diagrams for the beam in Figure 6-13.

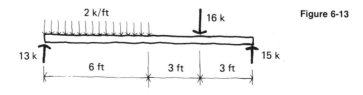

Figure 6-13

SOLUTION: Three separate sets of V and M equations will be used.

$$\Sigma F_y = 0$$
$$13 - 2x - V_x = 0$$

$(0 \leqslant x \leqslant 6) \quad V_x = 13 - 2x$

$$\Sigma M_x = 0$$

$$-13x + 2x\left(\frac{x}{2}\right) + M_x = 0$$

$(0 \leqslant x \leqslant 6) \quad M_x = 13x - x^2$

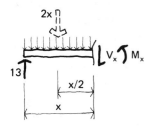

Figure 6-14 FBD for interval $(0 \leqslant x \leqslant 6)$.

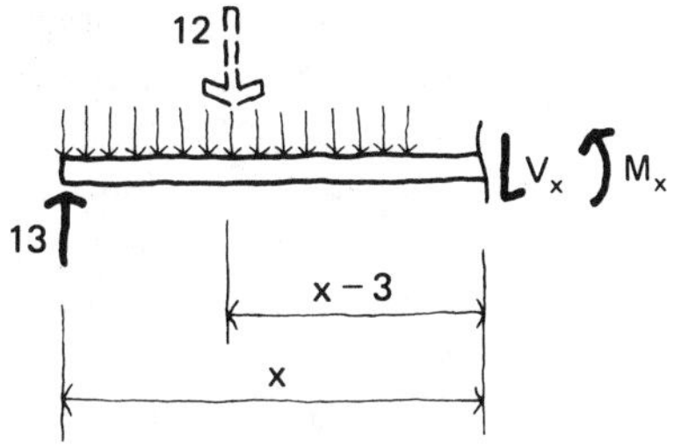

Figure 6-15 FBD for interval $(6 \leqslant x \leqslant 9)$.

$$\Sigma F_y = 0$$
$$13 - 12 - V_x = 0$$

$(6 \leqslant x \leqslant 9) \quad V_x = 1$

$$\Sigma M_x = 0$$

$$-13x + 12(x - 3) + M_x = 0$$

$(6 \leqslant x \leqslant 9) \quad M_x = 13x - 12x + 36$

Figure 6-16 FBD for interval $(9 \leqslant x \leqslant 12)$.

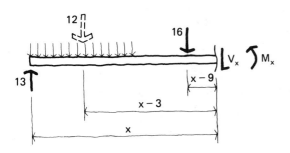

$$\Sigma F_y = 0$$

$$13 - 12 - 16 - V_x = 0$$

$$(9 \leqslant x \leqslant 12) \qquad V_x = -15$$

$$\Sigma M_x = 0$$

$$-13x + 12(x - 3) + 16(x - 9) + M_x = 0$$

$$(9 \leqslant x \leqslant 12) \qquad M_x = 13x - 28x + 180$$

Using these equations and values of x as needed, the diagrams of Figure 6-17 can be drawn. Notice that equations which contain the variable x, plot as straight horizontal lines; that equations containing x raised only to the first power plot as straight sloping lines, and that equations involving x^2 plot as parabolic curves. Note also that the points of curve change on the diagrams are common to two equations, and either equation may be used to find the ordinate value. Downward-acting uniform loads will result in moment curves that are concave downward, as verified by the value at $x = 3$ on the M diagram. Concentrated loads will always cause a sudden change in ordinate on the shear diagram and require two values of V at that section. (Actually, these two values of V, differing by the magnitude of the concentrated load, are a small distance apart because the load does occupy a short length of beam. In theory, however, we assume a true point load and the distance becomes infinitesimal.)

Using the guidelines given above, it is usually possible to construct shear and moment diagrams using only the values of x at the points of load change (i.e., the interval limits).

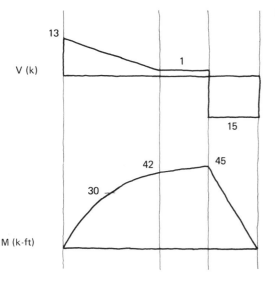

Figure 6-17 V and M diagrams for Example 6-3.

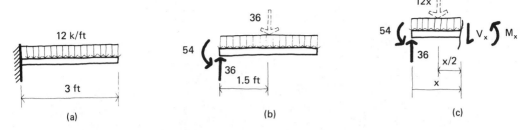

Figure 6-18   Cantilever beam with a uniform load.

Example 6-4

Construct the *V* and *M* diagrams for the beam in Figure 6-18(a).

SOLUTION: The external reactions are found in Figure 6-18(b) and the FBD needed to write the equations appears in Figure 6-18(c).

$$\Sigma F_y = 0$$
$$36 - 12x - V_x = 0$$
$$(0 \leq x \leq 3) \quad V_x = 36 - 12x$$

$$\Sigma M_x = 0$$
$$54 - 36x + 12x\left(\frac{x}{2}\right) + M_x = 0$$

$$(0 \leq x \leq 3) \quad M_x = -54 + 36x - 6x^2$$

This example serves well to illustrate the difference in the two sign conventions. In the FBD of Figure 6-18(c), the moment reaction is counterclockwise or plus in the $\Sigma M_x = 0$ statics equation. It causes tension in the top fiber, however (as the deflected shape of the beam would be concave downward) and is plotted as a negative ordinate on the moment diagram (Figure 6-19). Also notice that, just as a point load (or reaction) causes a sudden jump in the shear diagram, an externally applied moment

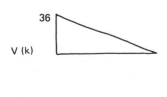

Figure 6-19   *V* and *M* diagrams for Example 6-4.

will cause a sudden jump in the moment diagram. *(Actually, this is the only way that such a change can occur in a moment diagram, which means that the moment will always be zero at the end of a beam unless there is an applied moment load or a wall reaction at that point.)*

As an exercise the student should turn the cantilever beam of Example 6-4 end for end and again construct the *V* and *M* diagrams. Since the origin is always kept at the left end, *x* will then equal zero at the free end and the two equations will change. The moment diagram remains all negative and is merely turned right about, as would be expected, but the shear diagram changes sign illustrating that the shear sense is not related to the beam's physical behavior.

PROBLEMS

6-1.   Construct the *V* and *M* diagrams for the simple beam in Figure 6-20.

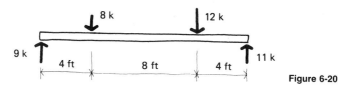

Figure 6-20

6-2.   Construct the *V* and *M* diagrams for the overhanging beam in Figure 6-21.

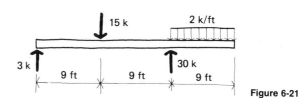

Figure 6-21

6-3.   Determine the reactions and construct the *V* and *M* diagrams for the beam in Figure 6-22.

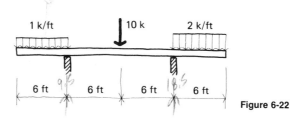

Figure 6-22

6-4.   Determine the reactions and construct the *V* and *M* diagrams for the cantilever beam in Figure 6-23.

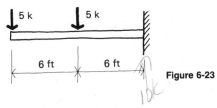

Figure 6-23

6-5. The beam in Figure 6-24 is partially restrained by its supports, resulting in applied couple loads at its ends. Determine the reactions and construct the *V* and *M* diagrams.

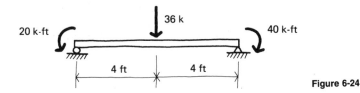

Figure 6-24

6-3 SIGNIFICANCE OF ZERO SHEAR

The examples and problems of the previous section had shear and moment diagrams that could be drawn by connecting the ordinates at the various points of curve change with either straight or curved lines. With the exception of Example 6-2, the maximum value of moment in each case occurred at one of these points. (For Example 6-2, the location of the maximum bending was obviously at midspan by symmetry.) For many loading situations, the point of maximum moment cannot be found so conveniently. However, it must be located before the magnitude of the maximum moment can be determined. Example 6-5 illustrates this situation.

Example 6-5

Construct the *V* and *M* diagrams for the partially loaded beam of Figure 6-25.

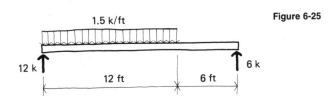

Figure 6-25

SOLUTION: The V and M equations for the two intervals will be

$$(0 \leqslant x \leqslant 12) \qquad V_x = 12 - 1.5x$$
$$(0 \leqslant x \leqslant 12) \qquad M_x = 12x - 0.75x^2$$
$$(12 \leqslant x \leqslant 18) \qquad V_x = -6$$
$$(12 \leqslant x \leqslant 18) \qquad M_x = 108 - 6x$$

The location of the point of maximum moment is not obvious from Figure 6-26. The distance x can be found rather easily, however, by remembering that the first derivative of an equation represents the slope. Borrowing the maxima-minima concept from calculus, we can find where the slope levels off to zero by taking dM_x/dx and setting it equal to zero. For the interval $0 \leqslant x \leqslant 12$, we have

$$M_x = 12x - 0.75x^2$$

and

$$\frac{dM_x}{dx} = 12 - 1.5x$$

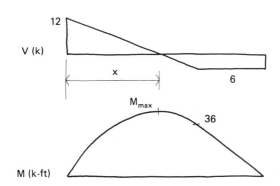

M (k-ft)   **Figure 6-26** V and M diagrams for Example 6-5.

Setting this expression to zero, we get

$$12 - 1.5x = 0$$
$$x = 8$$

By substitution, the value of the maximum moment is then

$$M_{x_{max}} = 12(8) - 0.75(8)^2$$
$$= 48 \text{ kip-ft}$$

It is most important to notice that the first derivative of the moment equation is equal to the *V* equation. Hence, when the moment becomes a maximum (or minimum) between points of curve change, it will do so where the shear is zero. This means that the value of *x* could have been determined from the slope of the *V* diagram. The slope is known as 1.5 kips per foot of length; therefore, to reduce the left-hand shear ordinate from 12 to 0 will require

$$\frac{12 \text{ kips}}{1.5 \text{ kips/ft}} = 8 \text{ ft}$$

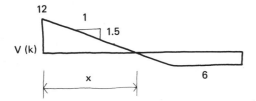

Figure 6-27 Similar triangles.

The relationships between the two diagrams will be explored more fully in the next section.

PROBLEMS

6-6. Construct the *V* and *M* diagrams for the beam in Figure 6-28.

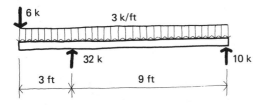

Figure 6-28 Overhanging beam.

6-7. What is the value of maximum moment in the beam of Figure 6-29? It carries a uniform load and is subjected to an applied moment at its left end.

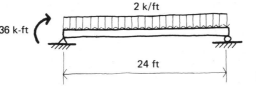

Figure 6-29

6-4   LOAD, SHEAR, AND MOMENT RELATIONSHIPS

Let us look further at what goes on inside a beam by studying the forces acting on a small length of the uniformly loaded simple span of Example 6-2. In Figure 6-30, we have added a load diagram which is nothing more than a plot of the transverse loads (including reactions) that act on the beam. Up loads are taken as positive and down loads as negative.

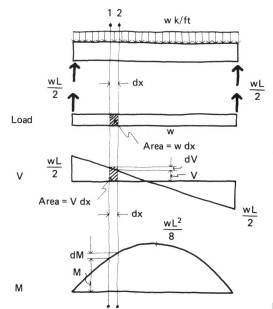

Figure 6-30   Load, shear, and moment diagrams.

Figure 6-31 shows a small elemental length of beam taken from between sections 1 and 2. This element must be in equilibrium under the forces shown; therefore,

$$\Sigma F_y = 0$$
$$V - w\,dx - (V - dV) = 0$$
$$dV = w\,dx \tag{6-1a}$$

or

$$w = \frac{dV}{dx} \tag{6-2a}$$

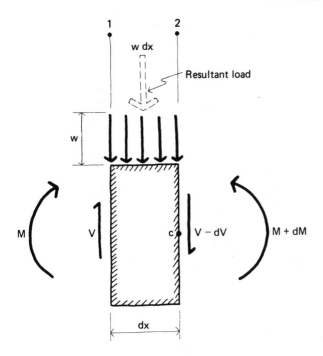

Figure 6-31   Free-body diagram.

Likewise, for rotational equilibrium, we can take moments about any point. Point c conveniently eliminates a force, and thus

$$\Sigma M_c = 0$$

$$M + dM + w \, dx\left(\frac{dx}{2}\right) - V \, dx - M = 0$$

Considering the third term to be small enough to be neglected gives us

$$dM = V \, dx \qquad\qquad (6\text{-}1b)$$

or

$$V = \frac{dM}{dx} \qquad\qquad (6\text{-}2b)$$

The four equations, (6-1) and (6-2), establish some very useful relationships among the load, shear, and moment diagrams. They enable us to construct the diagrams rapidly, without writing the shear and moment equations.

Look first at Equations (6-2) and let the distance dx in Figure 6-30 approach zero.

$$w = \frac{dV}{dx}$$ At any point along the length of the beam, the *ordinate* on the load diagram is equal to the *slope* of the shear diagram.

$$V = \frac{dM}{dx}$$ At any point along the length of the beam, the *ordinate* on the shear diagram is equal to the *slope* of the moment diagram.

This slope-ordinate relationship holds for both magnitude and sign. (A positive slope is one that is up and to the right; a negative slope is down and to the right.) Looking at the diagrams, we see that, in the left half of the beam, the ordinate of the V diagram and the slope of the M diagram are both positive and decreasing as we move from left to right. At midspan, the ordinate of the V diagram and the slope of the M diagram are both zero. Moving from left to right for the right half of the beam, we find that the ordinate of the V diagram and the slope of the M diagram are both negative and increasing.

Now look at Equations (6-1) and let the distance dx in Figure 6-30 be a small but finite interval.

$dV = w\,dx$ Over any beam length interval, the *net area* under the load curve is equal to the *change in ordinate* on the shear diagram.

$dM = V\,dx$ Over any beam length interval, the *net area* under the shear curve is equal to the *change in ordinate* on the moment diagram.

For the left half of the beam, the area under the shear curve is

$$A = \frac{1}{2}\left(\frac{wL}{2}\right)\left(\frac{L}{2}\right) = \frac{wL^2}{8}$$

This is the change in ordinate on the M diagram from $x = 0$ to $x = L/2$, that same interval. For the entire beam length, the *net* area under the shear curve is zero because the positive area equals the negative area. The change in ordinate on the M diagram from $x = 0$ to $x = L$ is zero, both points having a value of $M = 0$.

These relationships support and verify the point made previously concerning maximum moment and zero shear. We can also now state that, for a beam portion having no load ($w\,dx = 0$), the shear must be constant ($dV = 0$). Furthermore, if the moment is constant over a beam length ($dM = 0$), then no shear can exist ($V\,dx = 0$).

The reader should study the illustrations in Figure 6-32 to become familiar with the diagram relationships. It is recommended that Problems 6-1 through 6-7 be reworked for practice using the new techniques before attempting any new problems.

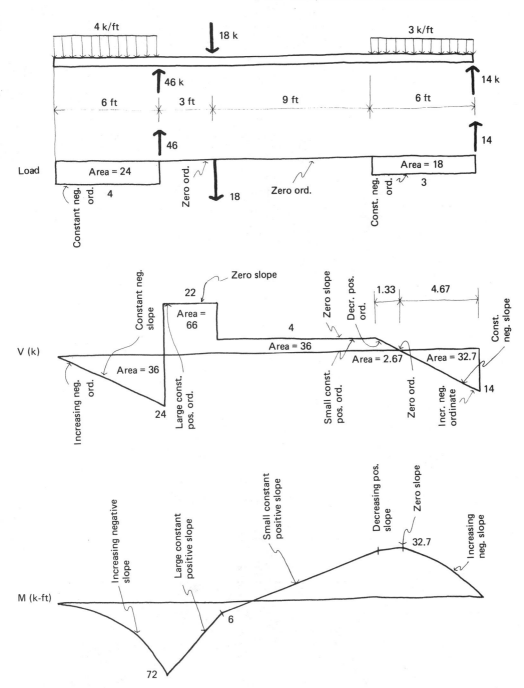

Figure 6-32 Diagram relationships.

Always remember to sketch the probable deflected shape and make sure it can be rationalized with the moment diagram in each case. Many careless errors can be found or prevented this way.

PROBLEMS

6-8. Construct the *V* and *M* diagrams for the long-span girder of Figure 6-33.

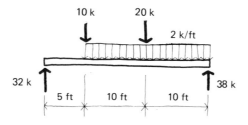

Figure 6-33

6-9. Construct the *V* and *M* diagrams for the beam in Figure 6-34.

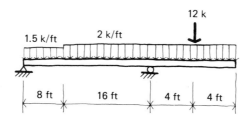

Figure 6-34

6-10. Construct the *V* and *M* diagrams for the cantilever beam in Figure 6-35.

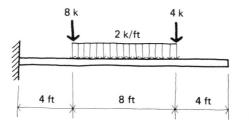

Figure 6-35

6-11. Construct the *V* and *M* diagrams for the beam in Figure 6-36. Pinned connections may be assumed.

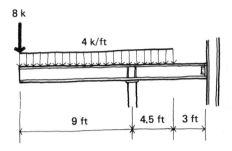

Figure 6-36

6-12. A simple beam of length L (ft) supporting a uniform load of w (kip-ft) has a midspan moment of $wL^2/8$ (kip-ft). How much moment, M (in terms of w and L), should be applied to the ends of the beam in Figure 6-37 to reduce that midspan moment by a factor of three?

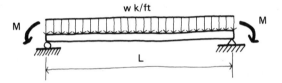

Figure 6-37

6-13. Construct the V and M diagrams for the hinged beam of Figure 6-38. (*Hint*: Make sure that the moment diagram goes to zero at the hinge.)

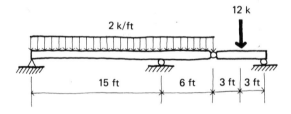

Figure 6-38 Hinged beam.

7

flexural stresses

7-1 INTRODUCTION

Chapter 1 makes reference to the structural inefficiency of bending as a means to carry load. Compared to axial tension and compression, bending generates much higher stresses in members, because the fibers of material are not stressed uniformly. We know from experience that if we want to break something in half, it is much easier to break it in bending than to try to pull it apart in tension or buckle it in compression.

Although it might be nice (for the sake of structural efficiency) to eliminate bending stresses in our structures, clearly this is not possible. To attempt to carry all loads in direct tension or compression would lead to some very awkward configurations and/or very small spans. The fact is that true structural efficiency, in many cases, would result in a false economy, because of the resulting spatial or architectural inefficiencies. (It is also true that bending action takes place in most structural elements anyway, even those designed and configured to carry loads by other means. As soon as we build stiffness into a structure, by thickness or shape (moment of inertia), it will have a tendency to carry load by bending.)

Many functions require clear spans in the short-to-moderate range of 15 to 60 ft, and most often some form of beam construction is preferable to arches, cables, vaults, folded plates, or other ''more efficient'' structures. A post-and-beam system will usually provide more usable building volume than the more form-resistant structures. For anything but the longer spans, beams will give the largest span/depth ratio, enabling the designer to reduce the space between the ceiling and the finish floor above. This can be essential in high-rise buildings.

Beams are also relatively insensitive to the type and location of loads they can receive. For example, trusses will only take concentrated loads at the panel points, and vaults and domes are quite unsuitable for either concentrated or line loads. Although it is true that bending action results in high stresses and deflections, it is equally true that strong, stiff structural materials are both readily available and inexpensive, particularly when compared to some nonstructural building materials and labor.

A beam can be loosely defined as any structural element that carries transverse loads and has two of its dimensions much less than the third. Most beams are straight and of constant cross section, but some are curved or angled and some have varying cross sections. All beams develop two kinds of stress: flexural (sometimes called bending), which is a normal stress, and shearing, which is tangential. All beams also deflect under load, and these three items—flexural stress, shearing stress, and deflection—are the parameters by which we determine the required size and/or shape of the cross section. Of these, flexural stress governs most frequently. The remainder of this Chapter and Chapter 8 will be devoted to analyzing and understanding beam stresses. Beam deflection is covered in Chapter 9.

7-2 FLEXURAL STRAIN

Beams bend under load such that transverse sections remain plane as represented by the lines on the beam in Figure 7-1(b). There is a compression zone and a tension zone, which are separated by a horizontal neutral plane. The neutral plane (called the *neutral axis*) is located at the centroid of the cross section and the beam fibers

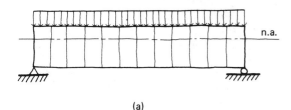

(a)

Figure 7-1 Flexural strain.

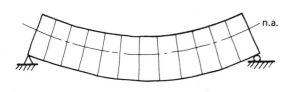

(b)

are squeezed or stretched in direct proportion to their distance from this neutral axis. The top and bottom fibers, or extreme fibers, undergo the most strain and do the most work, while those close to the neutral axis have strain levels near zero and are least effective. A fiber located halfway between the neutral axis and the top or bottom edge of a beam will have half the strain of a fiber located at that edge. (Verify this by making a small beam of an easy-to-bend material, such as polyurethane foam and drawing parallel lines on one of its sides.) It is important to understand the linearity of bending strain and the corresponding stresses that develop, and this subject is fully examined in Appendix A.

The correct location of the neutral axis, critical to the proper understanding of flexural stresses, was theorized in 1713 by Antoine Parent (1666–1716), a French scientist. However, it remained largely unknown until the extensive work by Claude L. M. H. Navier, also French, in 1826.

7-3   FLEXURAL STRESS

External bending loads are resisted by internal stresses that build up in the beam fibers. These stresses are directly related to flexural strain by the stiffness of the material. Their direction is always normal to the transverse section. Assuming that the beam is made of a reasonably homogeneous material, the stresses will vary as the strains do (i.e., linear with the distance from the neutral axis). Bending stress distributions for two different beam shapes are illustrated in Figure 7-2. Because of symmetry, the top and bottom fiber stresses in a rectangular section will be equal in magnitude. This is not true for the T shape, where higher stresses exist in the lower fibers of the stem than in those of the flange because of their greater distance from the neutral axis. In all cases, the stresses above and below the neutral axis will be opposite in sense.

Flexural stresses are a direct response to bending moments and therefore will vary over the length of a beam as well as with the distance from the neutral axis. The general flexure formula is

$$f_y = \frac{My}{I} \tag{7-1}$$

where f_y = flexural stress at fiber level y (psi or ksi)

M = bending moment at the transverse section being examined (lb-in. or kip-in.)

y = vertical distance from neutral axis to level y (in.)

I = moment of inertia of the cross section with respect to the neutral (centroidal) axis (in^4)

(*Note*: The examples and problems in this book assume that bending occurs about the strong axis and, therefore, I takes the value of I_x.)

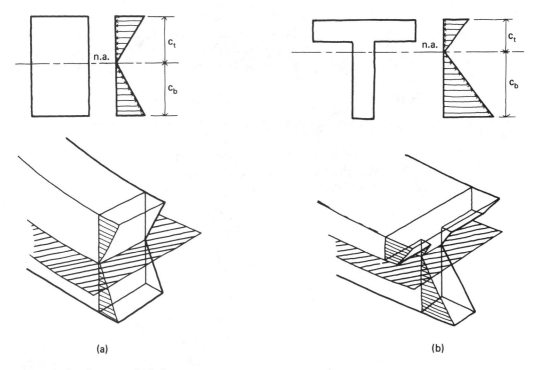

(a)   **(b)**

Figure 7-2 Bending stress distribution.

When we are interested only in the maximum bending stresses at a given transverse section, we can use

$$f_b = \frac{Mc}{I} \tag{7-1a}$$

where f_b = extreme fiber bending stress (psi or ksi)
$\quad\;\; c$ = distance to extreme fiber (in.)

[In Equation (7-1a), c will have the value illustrated as distance c_t or c_b in Figure 7-2, and f_b will then be a top fiber stress or a bottom fiber stress, respectively.]

The derivation of these formulas and certain restrictions concerning their use are given in Appendix A. The reader will notice that the sense of the bending stress at a given point in a beam depends upon the sense of the bending moment and whether the point is above or below the neutral axis.

(In some of the examples and problems that follow, the effects of member self-weight as part of the dead load have been ignored. Although this is not recommended as sound engineering practice, the writer's own experience indicates that this component of the load seldom controls the member size in short-span building structures of timber or steel. Larger spans in these materials and all reinforced concrete beams, however, have significant self-weights, which cannot be ignored.)

Example 7-1

The wood joist in Figure 7-3 is nominal 2 × 10 (1½ × 9¼).

(a) Determine the maximum bending stress.
(b) Determine the stress due to bending at a point 4.5 ft in from one of the ends and 3 in. below the top edge.

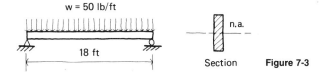

Figure 7-3

SOLUTION:

(a) By symmetry the neutral axis (n.a.) is located at mid-depth. The maximum bending stress will occur where the moment is a maximum and will be compressive in the top fiber and tensile in the bottom fiber. From Appendix I, $I = 98.9$ in^4.

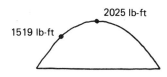

Figure 7-4 Moment diagram for the beam in Figure 7-3.

$$f_b = \frac{Mc}{I}$$

$$= \frac{2025 \text{ lb-ft } (12 \text{ in./ft}) (4.625 \text{ in.})}{98.9 \text{ in}^4}$$

$$f_{b_{top}} = 1140 \text{ psi compression}$$

$$f_{b_{bottom}} = 1140 \text{ psi tension}$$

(b) The point is above the neutral axis, and in the presence of positive moment the stress will be compressive.

$$f_y = \frac{My}{I}$$
$$= \frac{1519 \text{ lb-ft (12 in./ft) (1.625 in.)}}{98.9 \text{ in}^4}$$
$$= 300 \text{ psi compression}$$

Example 7-2

Determine the maximum bending stress in the beam of Figure 7-5.

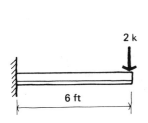

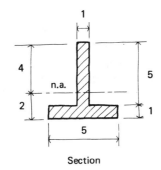

Section

Figure 7-5   Inverted T-beam used as a cantilever.

SOLUTION: The maximum bending stress will be at the cross section where the moment maximizes and at the fiber level farthest from the n.a. In this case, these conditions are met at the top fiber, where the beam enters the wall. The stress will be tensile.

12 k-ft

Figure 7-6   Moment diagram for the beam in Figure 7-5.

Using the parallel axis theorem, we get

$$I = 33.3 \text{ in}^4$$

The maximum moment is 12 kip-ft.

$$f_{b_{\text{top}}} = \frac{Mc}{I}$$
$$= \frac{12 \text{ kip-ft (12 in./ft) (4 in.)}}{33.3 \text{ in}^4}$$
$$= 17.3 \text{ ksi}$$

Example 7-3

A steel W12 × 16 is used for the beam in Figure 7-7. Determine the maximum tensile and compressive stresses due to bending.

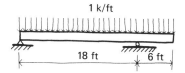

Figure 7-7   Overhanging beam with uniform load.

SOLUTION: As with the rectangle, the symmetrical W shape has its c distance to the top fiber equal to the c distance to the bottom fiber, so that (for any given section) the top and bottom stresses are equal in magnitude. This means that the maximum tensile and compressive stresses must both occur where the moment is a maximum, in this case, 32 kip-ft. From Appendix J, $I = 103$ in^4 and $d = 11.99$ in.

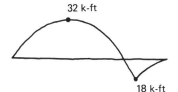

18 k-ft   **Figure 7-8**   Moment diagram for the beam in Figure 7-7.

$$f_{b_{\text{top}}} = \frac{Mc}{I}$$
$$= \frac{32 \text{ kip-ft } (12 \text{ in./ft}) (6 \text{ in.})}{103 \text{ in}^4}$$
$$= 22.4 \text{ ksi compression}$$

and

$$f_{b_{\text{bottom}}} = \frac{Mc}{I}$$
$$= \frac{32 \text{ kip-ft } (12 \text{ in./ft}) (6 \text{ in.})}{103 \text{ in}^4}$$
$$= 22.4 \text{ ksi tension}$$

Example 7-4

To save construction depth with a precast plank system, a structural T is used in lieu of the W shape in Example 7-3. Its properties are given in Figure 7-9. Determine the maximum tensile and compressive stresses due to bending.

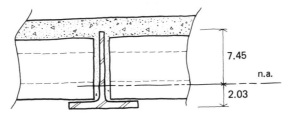

I = 119 in.⁴

Figure 7-9

SOLUTION: The structural T is not symmetrical with respect to its extreme fiber distances, and the possibility exists that one of the maximum stresses (tensile or compressive) may occur where the moment is not at its maximum absolute value. The compression stress will maximize at fibers 1 and 2 in Figure 7-10, and the tension stress will maximize at fibers 3 and 4. For illustrative purposes we shall examine all four points in this example. (The moment diagram from Example 7-3 is, of course, still valid.)

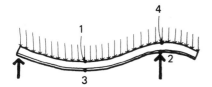

Figure 7-10   Deflected shape of structural T-beam.

For the maximum positive moment section,

$$f_{b_1} = \frac{Mc}{I}$$
$$= \frac{32 \text{ kip-ft } (12 \text{ in./ft}) (7.45 \text{ in.})}{119 \text{ in}^4}$$
$$= 24.0 \text{ ksi compression}$$

and

$$f_{b_3} = \frac{Mc}{I}$$
$$= \frac{32 \text{ kip-ft } (12 \text{ in./ft}) (2.03 \text{ in.})}{119 \text{ in}^4}$$
$$= 6.55 \text{ ksi tension}$$

For the maximum negative moment section,

$$f_{b_4} = \frac{Mc}{I}$$
$$= \frac{18 \text{ kip-ft } (12 \text{ in./ft}) (7.45 \text{ in.})}{119 \text{ in}^4}$$
$$= 13.5 \text{ ksi tension}$$

and

$$f_{b_2} = \frac{Mc}{I}$$
$$= \frac{18 \text{ kip-ft } (12 \text{ in./ft}) (2.03 \text{ in.})}{119 \text{ in}^4}$$
$$= 3.69 \text{ ksi compression}$$

A comparison of the four values just determined shows that the maximum compression occurs where the moment is maximum (i.e., at point 1). The tensile stress, however, maximizes at point 4, where the moment is only 18 kip-ft. This happens, of course, because of the different c distances involved in the asymmetrical T shape. Accordingly, when sections having dissimilar c distances are used for beams that have both positive and negative curvature, the largest bending stress of a specified sense will not necessarily occur where the moment is maximum. In order to determine the maximum bending stresses in such cases, one either has to check the stress levels at four points (as we did) or compare the ratio of M values to the ratio of c distances to ascertain where the equation Mc/I will maximize. What remains true, however, is that the *absolute* maximum value of bending stress (24.0 ksi in our example) will always occur at the section having the *absolute* maximum moment.

PROBLEMS

7-1. A W24 × 68 is used as a simple beam spanning 30 ft. It must carry a uniform load of 1 kip/ft and two concentrated loads at the third points of 10 kips each. Determine the maximum bending stress.

7-2. Douglas fir 2 × 12 joists span 17 ft between two bearing walls. Determine the permissible total uniform load in lb/ft. Use the repetitive member bending stress in Appendix H as the allowable stress.

7-3. Determine the maximum permissible simple span for the joists of Problem 7-2 if the total uniform load is 70 lb/ft. Assume that bending stress controls.

7-4. The built-up timber beam of Figure 7-11 is made of two 2 × 10 members enclosing a 2 × 6. The beams are spaced 8 ft apart in a direction normal to the page and must carry a floor load of 30 psf. Determine the maximum tensile and compressive stresses due to bending. (*Hint*: Each linear foot of beam must support 8 square feet of floor area.)

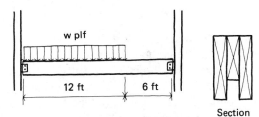

Section

Figure 7-11 Balcony beam simply supported by two columns. (The letters plf mean pounds per linear foot.)

7-5. A W21 × 50 is used for the overhanging beam of Figure 7-12. Assuming pinned connections, determine the magnitude and sense of the extreme fiber stresses
(a) at the section where the moment is maximum.
(b) under the left-hand concentrated load.

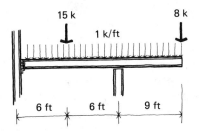

Figure 7-12

7-6. The 24-ft-long timber 4 × 12 beams of Figure 7-13 are spaced 4 ft on center. They must carry a total floor load of 40 psf and a wall load (from the roof) of 1 kip per running foot of wall. Determine the maximum bending stress.

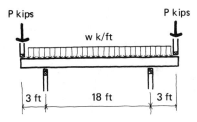

Figure 7-13 Section through a residential floor.

7-7. The steel beam of Figure 7-14 is composed of a W section and a channel welded together and has the *I* value given. Determine the maximum compressive and tensile stresses due to bending.

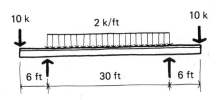

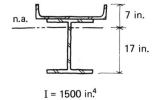

I = 1500 in.^{4} **Figure 7-14**

$C = 17$

7-4 SECTION MODULUS

The general formula for flexural stress, $f_b = Mc/I$, can be simplified slightly if we restrict our analyses to problems involving only extreme fiber stresses. (Since these are the stresses that usually control, this restriction is of little consequence.) Notice that the maximum flexural stress is really a function of only two items, the bending moment and the dimensions of the cross section. If we can combine the two cross-sectional factors c and I into one term, the general equation will be easier to use, particularly when applied in a design situation. The quantity I/c has been given the special name of section modulus and the symbol S. It is a measure of bending resistance which includes both the moment of inertia and the depth. Its units are length cubed. The formula for maximum flexural stress will then be

$$f_b = \frac{M}{S} \tag{7-2}$$

In a section that is symmetrical about the neutral axis, the c distances to the tensile and compressive fibers will be equal and the section modulus will have only one value. For a T shape or other unsymmetrical section, where the neutral axis is not at middepth, the larger c dimension should be used in S so that formula (7-2) will compute the larger of the two extreme fiber stresses. (It is probably just as easy to use the straight Mc/I formula for unsymmetrical shapes.)

For a rectangular section, S can be stated in terms of the width and depth, bypassing the I computation.

$$\begin{aligned} S &= \frac{I}{c} \\ &= \frac{bd^3/12}{d/2} \\ &= \frac{bd^2}{6} \end{aligned} \tag{7-3}$$

Section modulus values for the strong axes (S_x) of some common timber rectangles have been computed and are given in Appendix I. For selected steel shapes S_x values are listed in Appendix J. The examples and problems in this book all assume strong-axis bending.

Example 7-5

Determine the maximum bending stress in a 4 × 10 timber, which is used as a uniformly loaded cantilever 6 ft long. The total load is 300 plf.

SOLUTION: From Appendix I, $S = 49.9$ in^3.

$$f_b = \frac{M}{S}$$
$$= \frac{5400 \text{ lb-ft}(12 \text{ in./ft})}{49.9 \text{ in}^3}$$
$$= 1300 \text{ psi}$$

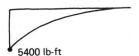

5400 lb-ft   **Figure 7-15**   Moment diagram for the beam of Example 7-5.

Example 7-6

The W36 × 260 in Figure 7-16 must carry three column loads of 75 kips each. Determine the maximum bending stress.

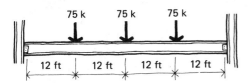

Figure 7-16   Simply supported steel beam.

SOLUTION: From Appendix J, $S = 953 \text{ in}^3$,

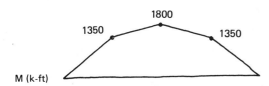

Figure 7-17   Moment diagram for the beam in Figure 7-16.

$$f_b = \frac{M}{S}$$
$$= \frac{1800 \text{ kip-ft}(12 \text{ in./ft})}{953 \text{ in}^3}$$
$$= 22.7 \text{ ksi}$$

This is very close to the allowable bending stress of 24 ksi for mild steel, and the size of the span indicates that the bending stress due to the dead weight of the beam itself should not be ignored. The self-weight is a uniform load that (in this case) causes a maximum moment at midspan which should be added to the applied load moment at that point. The self-weight moment is

$$M_{s.w.} = \frac{wL^2}{8}$$

where w is the self-weight of the beam (plf or klf). In this case $w = 260$ plf or 0.260 klf. Therefore,

$$M_{s.w.} = \frac{0.260 \text{ kip/ft}(48 \text{ ft})^2}{8}$$
$$= 75 \text{ kip-ft}$$

The actual bending stress after inclusion of the self-weight moment will be, by ratio,

$$f_{b_{new}} = \frac{M + M_{s.w.}}{M}(f_b)$$
$$= \frac{1800 + 75}{1800}(22.7 \text{ ksi})$$
$$= 23.6 \text{ ksi}$$

This is still within the allowable of 24 ksi, so a larger beam will not be required.

Example 7-7

A hem-fir joist must span 15 ft and carry a total uniform load of 67 plf. Assuming that flexure controls the design, select the smallest adequate 2 × ? section. Use the allowable stress for repetitive members.

SOLUTION:

$$M = \frac{wL^2}{8}$$
$$= \frac{67 \text{ lb/ft}(15 \text{ ft})^2}{8}$$
$$= 1880 \text{ lb-ft}$$
$$S_{required} = S_r = \frac{M}{F_b}$$

The value of allowable stress F_b from Appendix H is 1150 psi.

$$S_r = \frac{1880 \text{ lb-ft}(12 \text{ in./ft})}{1150 \text{ psi}}$$
$$= 19.6 \text{ in}^3$$

From Appendix I, a 2 × 10 with a section modulus of 21.4 in³ is the smallest adequate size.

PROBLEMS

7-8. Determine the maximum bending stress in a W18 × 40 steel beam that carries a midspan concentrated load of 20 kips on a simple span of 24 ft.

7-9. A large beam simply spans 74 ft and carries an applied load of 1.5 klf. Assuming that an allowable bending stress of 24 ksi will control the beam size, select the lightest adequate W shape from those listed below. Include the effect of member self-weight.
  (a) W36 × 150, S = 504 in³
  (b) W36 × 160, S = 542 in³
  (c) W36 × 170, S = 580 in³
  (d) W36 × 182, S = 623 in³

7-10. Select Douglas fir 2 × ? floor joists for each of the following conditions. Assume that bending will control and that repetitive member stresses apply.
  (a) w = 67 plf, L = 18 ft
  (b) w = 67 plf, L = 15 ft
  (c) w = 53 plf, L = 18 ft
  (d) w = 53 plf, L = 12 ft

7-11. Ignoring any deflection limitation and other controlling factors, how far can a W36 × 300 span before the flexural stress from its own weight will reach an allowable value of 24 ksi?

7-12. The beam in Figure 7-18 is a hemlock member having an actual cross section of 4 in. by 10 in. Determine the flexural stress
  (a) at the right-hand support
  (b) under the point load.

(*Hint*: Use $S = bd^2/6$.)

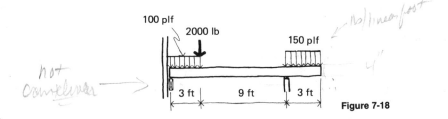

Figure 7-18

7-5 LATERAL BUCKLING AND STABILITY

Whenever a long, slender column is loaded in compression along its axis, it tends to deflect sideways, or *buckle*. This buckling phenomenon occurs even though the stresses remain well within the elastic range of the material. It occurs rapidly once a certain critical load is reached and is a function of the modulus of elasticity and cross-sectional shape rather than of material strength. (Elastic column buckling is

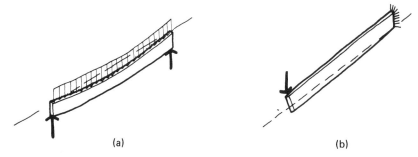

(a)   (b)

Figure 7-19 Lateral buckling.

discussed in Chapter 10.) This same behavior occurs in the compression zones of long slender beams.

Whenever compressive stresses exist over a length of beam, such as in the top of a simple beam or along the bottom of a cantilever, there exists a tendency for the compressive fibers to buckle laterally or "get out of the way of the compressive forces" (Figure 7-19). It makes no difference whether the loads are applied from above or below. The buckling is caused by the horizontal force resultant of the internal moment couple, not by the fact that loads push downward from above. Even though the tension fibers tend to remain straight, the section undergoes a rotation or twisting action, which reduces both the effective depth and the moment of inertia.

The examples and problems presented earlier all assumed that lateral buckling was not a factor or was prevented from happening in some manner. Certain beams are inherently stable against any lateral buckling tendency by virtue of cross-sectional shape. For example, a rectangle with a width greater than its depth and loaded vertically in a plane of symmetry will have no lateral stability problem. A wide-flange beam having a compression flange that is both wide and thick, so as to provide a resistance to bending in a horizontal plane, will have considerable resistance to buckling.

A beam that is not laterally stiff in cross section must be braced every so often along its compressive side in order to develop its full moment capacity. Sections not so braced or laterally supported by secondary members will fail prematurely (or at best be unsafe in terms of maintaining a proper factor of safety). Sometimes such lateral bracing occurs naturally because of other design considerations. The plywood subfloor nailed (and frequently glued) to the tops of wood joists of simple residential spans provides excellent lateral support. Open web bar joists, with their ends welded to the top flanges of the beams that carry them, provide lateral bracing for those flanges. Other situations, such as the overhanging beams of Figure 7-20, require specific bracing elements. In this case, the four beams are tied together by the spandrel channels at their ends and one bay has X-bracing (of rods or angles) connecting the critical compression flanges.

Reinforced concrete beams usually have cross-sectional dimensions such that lateral buckling is not a consideration. As mentioned previously, small timber joists

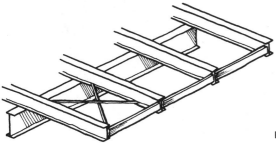

Figure 7-20 Lateral stability for overhanging steel beams.

and beams almost always have adequate lateral support for their top edges provided by the attached floor deck and required bridging. Similarly, the overhangs involved in small-scale timber construction are usually short so that the lateral stability of the underside is not a critical concern. Larger solid-sawn timber sections and glued-laminated beams, however, can easily have lateral support problems. The American Institute of Timber Construction (AITC) provides equations so that designers can compute reduced allowable stresses when adequate lateral support cannot be provided.

The issue of lateral stability occurs more frequently when designing with steel than with other materials. Its inherent strength means smaller sections and because of relatively high material costs, such sections tend to be efficiently configured for bending (i.e., deep and narrow). The American Institute of Steel Construction (AISC) has developed equations to determine the reduced bending capacities for members with inadequate lateral support and provides a series of graphs as design aids.

In a few design situations it may become desirable not to use full moment capacity of a section but rather to use reduced allowable stresses (calling for lower loads or larger members) in order to maintain the same margin of safety. An excellent illustration of this approach is found in Crown Hall (on the Illinois Institute of Technology campus) by Mies van der Rohe (see Figure 7-21). Here the large clear-span steel plate girders that frame the roof are exposed, with the roof deck attached to the bottom or tension flange.

Figure 7-21 Concept sketch of Crown Hall.

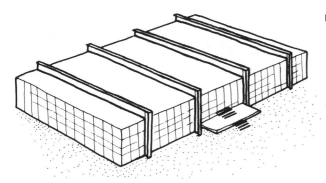

Clearly, the architect desired a strong statement of the horizontal structure, achieved by exposing these girders, and was similarly not concerned by the increase in their size required by the absence of lateral support elements.

The proper analysis of beams that lack lateral support has not been included in this basic text and is more properly treated in a context involving applied analysis and design procedures for specific materials.

8

shearing stresses

8-1 NATURE OF SHEARING STRESS

As introduced in Section 4-1, shearing stresses are tangential stresses that act parallel to the planes which they stress. Figure 8-1 shows how the shearing force in a beam provides shearing stresses on both vertical and horizontal planes within the beam. The two vertical stresses, f_v, must be equal in magnitude and opposite in sense to ensure vertical equilibrium. However, under the action of those two stresses alone, the element would rotate in a clockwise manner. Clearly, this couple must be negated by the action of another couple, shown as the dashed arrows. If the small element is taken as a differential one, the magnitude of the horizontal stresses must also have the value f_v. This principle is sometimes phrased as ''cross-shears are equal.'' In other words, a shearing stress cannot exist on an element without a like stress located $90°$ around the corner.

8-2 DIAGONAL TENSION AND COMPRESSION

Shearing stresses also create tensile and compressive stresses. The square element of Figure 8-2(a) is being acted upon by four shearing stresses as just explained. The stressed element will appear as in Figure 8-2(b) as it deforms, developing a tensile stress along a line from a to b and compressive stress along a line joining c and d. In the absence of any other stresses acting on the element, the lines of tension and compression will be oriented at $45°$ to the original shear planes.

164

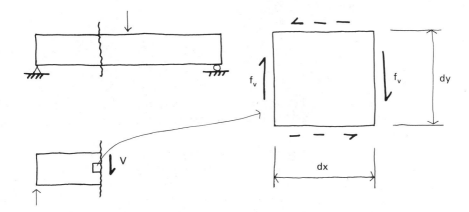

Figure 8-1 Development of shearing stresses in a beam.

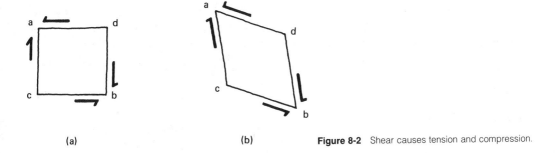

(a)   (b)   **Figure 8-2** Shear causes tension and compression.

If the thickness of the element is designated as dz, then an equation of equilibrium in the ab direction can be used to solve for the magnitude of f_t, the tensile stress. Referring to Figure 8-3,

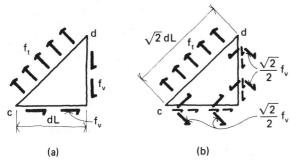

Figure 8-3 Free-body cut through cd, showing stresses.

(a)   (b)

$$\Sigma F_{ab} = 0$$

$$f_t(\sqrt{2})dL(dz) - 2\left[\frac{\sqrt{2}}{2}(f_v)\right]dL(dz) = 0$$

$$f_t = f_v$$

which illustrates that the diagonal tension developed by shearing stress is equal to the shearing stress itself. A similar proof could be made for the diagonal compression stress in the *cd* direction.

It is important to note that a material which is weak in either tension or compression will also be effectively weak in shear. Thus, it was explained in Chapter 5 that concrete is weak in shear because of its lack of strength in tension. Concrete beams are strengthened by specially placed reinforcing bars (called *stirrups*) to prevent diagonal tension cracking. Figure 8-4 shows potential diagonal tension cracks being crossed by several stirrups. (Stirrups are placed vertically rather than normal to the potential cracks for reasons of construction ease.)

Some deep steel girders have relatively thin webs, which tend to buckle in compression along 45° lines, as shown happening on the left-hand beam of Figure 8-5. Vertical plate stiffeners shown welded in place on the right-hand sketch constitute one way to prevent such failure. The reader can "see" diagonal buckling as it occurs in a thin member by applying shearing forces along the opposite edges of a piece of paper.

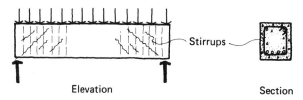

Elevation   Section

Figure 8-4 Concrete beam reinforced for shear.

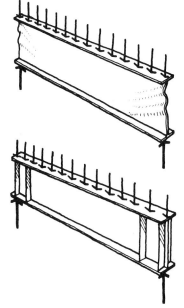

Figure 8-5 Adding stiffeners to a steel girder for shear.

8-3 BASIC HORIZONTAL SHEARING STRESS EQUATION

In some ways it is easier to visualize shearing stresses acting on horizontal planes than upon vertical ones. For example, if you make a beam by laying several planks flatwise on top of one another, there would exist horizontal slippage planes as shown in Figure 8-6. As the top fibers of each plank get shorter in compression, they have to "slip past" the bottom fibers of the plank above. The bottom fibers, in each case, are themselves getting longer because of the bending tensile strain. Now if we glued all the planks together, so as to simulate a solid one-piece cross section, there would be less deflection, and horizontal shearing stresses would develop in the glue planes. These same stresses occur in solid pieces, of course, and are particularly important in the design of wood beams because most softwood shears rather easily parallel to the grain. (Examine the values given in Appendix H and see that this is reflected in the relative magnitudes of the allowable stresses. The horizontal shearing stress value F_v is quite low.)

There is a close relationship between flexural stress and shearing stress. Clearly, the slippage deformations of Figure 8-6 would not take place in the absence of bending. Indeed, the derivation of the general shearing stress formula in Appendix B proves that such stresses are caused by the *change* in moment from one beam section to the next. This is also implied in Chapter 6, where it is stated that the magnitude of the ordinate on the V diagram (the shear force) is equal to the slope of the moment diagram. Zero shear can only exist when the slope of the moment diagram is zero.

Since shearing stresses must exist on all four planes of an element in order to exist at all, it follows that shearing stresses will be zero at the top and bottom edges of a beam where there is no material present to provide one of the four stresses. Unlike flexural stresses, which maximize at the extreme top and bottom fibers of a section, shearing stresses tend to maximize near the center of a beam cross section and go to zero at the extreme fibers. The general equation for horizontal (or vertical) shearing stress in beams is

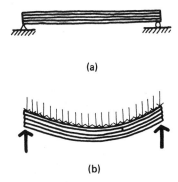

(a)

(b)

Figure 8-6 Beam made of planks.

$$f_v = \frac{VQ}{Ib} \tag{8-1}$$

where f_v = shearing stress (psi or ksi)

V = vertical shear force at the transverse section being examined (lb or kips)

Q = statical moment of that area of cross section between the horizontal plane under investigation and the near edge of the beam, taken with respect to the neutral axis (in^3)

I = moment of inertia of the cross section with respect to the neutral axis (in^4)

b = width of cross section at the horizontal plane under investigation (in.)

The term represented by Q in the formula is not nearly so complicated as its written definition implies. Q is really nothing more than a shape factor that represents how bending forces (which cause the shear) are distributed with respect to the neutral axis. Bending stress is linear, but bending force is a function of the stressed area as well and is not linear. As illustrated in Appendix B, this will cause the shear stress in beams to vary as a square function, or parabolically, over the depth of the cross section.A few examples will illustrate this more clearly.

Example 8-1

The 2 × 10 wood joist of Example 7-1 has the shear diagram shown in Figure 8-7. Determine the distribution of horizontal shearing stresses on a transverse section just to the inside of either support.

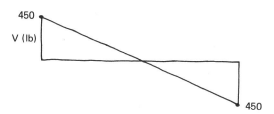

Figure 8-7 Shear diagram for wood joist in Example 7-1.

SOLUTION: The shearing force V has its maximum value of 450 lb at these two locations. For a given transverse section, V and I are constants and the shearing stress varies with the value of the ratio Q/b. For this cross section b is a constant also, so f_v will vary over the depth directly with the value of Q.

At the neutral axis, using Figure 8-8(a),

$$Q = 4.62(1.5)(2.31)$$
$$= 16 \text{ in}^3$$

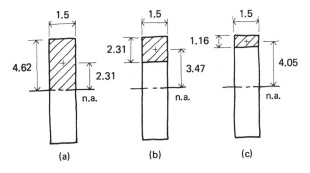

Figure 8-8 Determination of Q.

and the shearing stress will be

$$f_{v_{n.a.}} = \frac{VQ}{Ib}$$

$$= \frac{450 \text{ lb}(16 \text{ in}^3)}{98.9 \text{ in}^4(1.5 \text{ in.})}$$

$$= 48 \text{ psi}$$

Halfway to the edge of the section, at the $d/4$ level, Q is obtained using Figure 8-8(b).

$$Q = 2.31(1.5)(3.47)$$
$$= 12 \text{ in}^3$$

The shearing stress is

$$f_{v_{d/4}} = \frac{VQ}{Ib}$$

$$= \frac{450 \text{ lb}(12 \text{ in}^3)}{98.9 \text{ in}^4(1.5 \text{ in.})}$$

$$= 36 \text{ psi}$$

At the $d/8$ level, where Q is obtained using Figure 8-8(c),

$$Q = 1.16(1.5)(4.05)$$
$$= 7 \text{ in}^3$$
$$f_{v_{d/8}} = \frac{VQ}{Ib}$$

$$= \frac{450 \text{ lb}(7 \text{ in}^3)}{98.9 \text{ in}^4(1.5 \text{ in.})}$$

$$= 21 \text{ psi}$$

The values will be identical at symmetrical levels below the neutral axis, and a plot of the shearing stress is given in Figure 8-9. The distribution would look the same at the other transverse sections of the beam, but the values would be less because of the decrease in V.

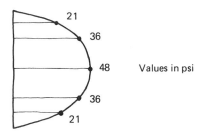

Figure 8-9   Shearing stress distribution for the rectangle of Example 8-1.

Example 8-2

Determine the shearing stress distribution for the beam in Figure 8-10.

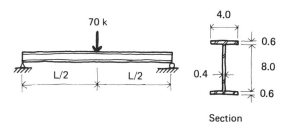

Figure 8-10

SOLUTION: V is constant for this beam at 35 kips. I can be determined as 106 in⁴. Using Figure 8-11, Q at level 1 (the neutral axis) is found to be

$$Q = 4(0.4)(2) + 0.6(4)(4.3)$$
$$= 13.5 \text{ in}^3$$

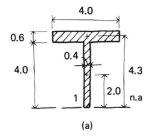

(a)

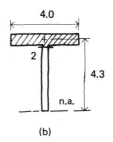

(b)

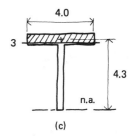

(c)

Figure 8-11

and

$$f_{v_1} = \frac{VQ}{Ib}$$

$$= \frac{35 \text{ kips}(13.5 \text{ in}^3)}{106 \text{ in}^4(0.4 \text{ in.})}$$

$$= 11.1 \text{ ksi}$$

At level 2, b remains the same but

$$Q = 0.6(4)(4.3)$$
$$= 10.3 \text{ in}^3$$

and

$$f_{v_2} = \frac{VQ}{Ib}$$

$$= \frac{35 \text{ kips}(10.3 \text{ in}^3)}{106 \text{ in}^4(0.4 \text{ in.})}$$

$$= 8.5 \text{ ksi}$$

At level 3, just inside the flange, b takes a sharp increase, which will cause a drop in the shearing stress. Q has the same value as at level 2.

$$f_{v_3} = \frac{VQ}{Ib}$$

$$= \frac{35 \text{ kips}(10.3 \text{ in}^3)}{106 \text{ in}^4(4.0 \text{ in.})}$$

$$= 0.85 \text{ ksi}$$

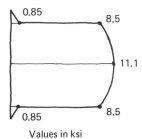

Values in ksi

Figure 8-12   Shearing stress distribution for the wide-flange shape of Example 8-2.

The final distribution is shown in Figure 8-12. It varies according to the ratio Q/b.

Example 8-3

Determine the plane of maximum shear stress for the cross section in Figure 8-13.

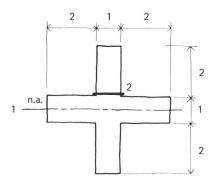

Figure 8-13   Cruciform beam cross section.

SOLUTION: The stress will maximize at the plane where the ratio Q/b maximizes. Q will always be a maximum at the neutral axis; therefore,

$$\left(\frac{Q}{b}\right)_1 = \frac{0.5(5)(0.25) + 2(1)(1.5)}{5}$$

$$= 0.725$$

The other possible place for Q/b to reach its largest value would be at the junction of the two rectangles, plane 2.

$$\left(\frac{Q}{b}\right)_2 = \frac{2(1)(1.5)}{1}$$

$$= 3.0$$

Figure 8-14 Qualitative shearing stress distribution for the cruciform shape of Example 8-3.

Clearly, the shearing stress will maximize at this junction. A distribution plot would appear as in Figure 8-14.

PROBLEMS

8-1. Determine the shearing stress distribution for the beam in Figure 7-5. Give the values in 1-in. increments of depth.

8-2. Show by taking successive trial planes (or by the calculus) that the shearing stress of a triangular cross section, such as that of Figure 8-15, maximizes at $h/2$.

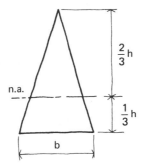

Figure 8-15

8-3. A Douglas fir rectangular timber 11½ in. deep is used as a simple beam spanning 12 ft. If the total load is a uniform 300 plf, determine the required safe minimum width.

8-4. Locate the plane of maximum shearing stress for the channel section of Figure 8-16.

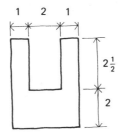

Figure 8-16 Channel shape.

8-4 HORIZONTAL SHEARING STRESSES IN TIMBER BEAMS

The general expression for shearing stress due to bending is $f_v = VQ/Ib$, as presented in Section 8-3. For a rectangular section, as most timber beams are, the maximum value of shearing stress occurs at the neutral axis. Since Q and I can both be expressed in terms of b and d for a rectangular section, a simpler expression for this maximum f_v can be developed.

$$f_{v_{max}} = \frac{VQ_{max}}{Ib}$$

$$Q_{max} = b\left(\frac{d}{2}\right)\left(\frac{d}{4}\right)$$

$$= \frac{bd^2}{8}$$

$$I = \frac{bd^3}{12}$$

$$= \frac{Vbd^2(12)}{bd^3(8)b}$$

$$bd = A$$

$$= \frac{3V}{2A} \tag{8-2}$$

From Equation (8-2), the maximum shearing stress is 50% larger than the average value, which can be represented by a rectangular block. The total shearing force resistance or shear capacity is stress times area, and this is represented as a volume in Figure 8-17. The area under a parabola is ⅔ the base times the height (see Appendix F), so its altitude must be 50% greater to achieve a "stress volume" equal to a rectangular volume.

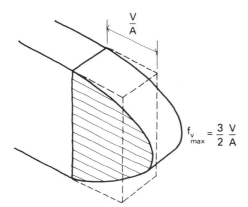

Figure 8-17   Shearing stress distribution.

Example 8-4

A hem-fir 4 × 6 beam spans 10 ft and supports a uniform load of 180 plf. Is it adequate in shear?

SOLUTION: The maximum shear force will be equal to one of the reactions.

$$V = \frac{wL}{2}$$

$$= \frac{180 \text{ lb/ft}(10 \text{ ft})}{2}$$

$$= 900 \text{ lb}$$

$$f_v = \frac{3V}{2A}$$ $A = 19.25 \text{ in}^2$

$$= \frac{3(900 \text{ lb})}{2(19.25 \text{ in}^2)}$$

$$= 70 \text{ psi}$$

From Appendix H the allowable shearing stress for hem-fir is 75 psi, so the section is adequate in shear.

PROBLEMS

8-5. A uniformly loaded southern pine joist must span 16 ft and carry a total load of 67 plf. Will a 2 × 10 section be OK in shear?

8-6. The beam in Figure 8-18 is a doubled 2 × 8. Compute the maximum shearing stress caused by the five concentrated loads.

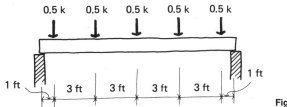

0.5 k   0.5 k   0.5 k   0.5 k   0.5 k

1 ft

1 ft   3 ft   3 ft   3 ft   3 ft

Figure 8-18

8-7. Compute the maximum shearing stress in the overhanging 2 × 12 joist in Figure 8-19.

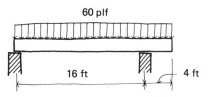

60 plf

16 ft   4 ft

Figure 8-19

8-5 HORIZONTAL SHEARING STRESSES IN STEEL BEAMS

As developed in Section 8-3, the distribution of shearing stresses in a W shape is as shown in Figure 8-20. Almost all of the shearing force is resisted by stresses in the web, and very little work is done by the flanges—the opposite, of course, being the case for flexural stresses. The calculation of the exact maximum stress magnitude using VQ/Ib can become difficult because of the presence of fillets where the flanges join the web. A high level of accuracy is even harder to achieve in channels or I shapes which have sloping flange surfaces. Accordingly, the American Institute of Steel Construction recommends the use of a much simpler approximate formula for the common steel shapes.

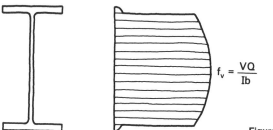

$$f_v = \frac{VQ}{Ib}$$

Figure 8-20 Shearing stress distribution in a wide-flange beam.

$$f_v = \frac{V}{th} \tag{8-3}$$

where t = web thickness (in.)
$\quad h$ = total beam depth (in.)

This formula gives the average unit shearing stress for the web over the full beam depth, ignoring any contribution of the flange projections (Figure 8-21). Depending upon the particular steel shape, this formula can be as much as 20% in error in the nonconservative direction. This means that when a shearing stress computed by Equation (8-3) gets to within 20% of the maximum allowable stress, the actual maximum stress (computable by VQ/Ib) might be exceeding the allowable by a small amount.

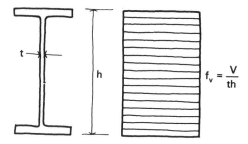

$$f_v = \frac{V}{th}$$

Figure 8-21 Simplified shearing stress distribution.

Fortunately, this low level of accuracy is seldom a problem for two reasons:

1. Structural steel is very strong in shear.
2. Most beams and girders in buildings, unlike those in some machines, have very low shearing stresses.

A rolled steel beam has to be very short and very heavily loaded, or have a large concentrated load adjacent to a support, in order for shear to control. In determining the size of a steel beam, flexural stresses will usually govern. Excessive deflection will occasionally dictate the use of a larger section, but shear will almost never govern the design.

When shearing stresses do become excessive, steel beams do not fail by ripping along the neutral axis as might happen in wood. Rather, it is the compression buckling of the relatively thin web which constitutes a shear failure. This can be diagonal buckling, as discussed in Section 8-2, or a type of vertical buckling, illustrated in Figure 8-22. The AISC has provided several design formulas for determining when extra bearing area must be provided at concentrated loads or when web stiffeners are needed to prevent such failures (Figure 8-23).

Most beams of normal depth seldom present any major problems, and detailed design considerations will not be given here. A word of caution is given with respect to large built-up plate girders, however. Such sections usually have deep thin webs and are particularly susceptible to buckling action. For these beams, shear can be a determining factor in the overall structural design. The reader may wish to consult the *Manual of Steel Construction* for further material on web buckling.

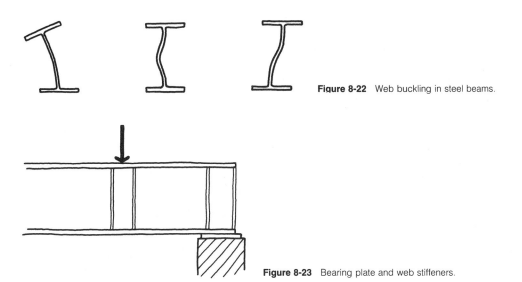

Figure 8-22 Web buckling in steel beams.

Figure 8-23 Bearing plate and web stiffeners.

Example 8-5

Determine the average shearing stress for the W16 × 57 in Figure 8-24 if V = 25 kips.

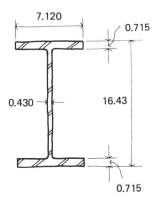

Figure 8-24

SOLUTION:

$$f_v = \frac{V}{th}$$

$$= \frac{25 \text{ kips}}{0.430 \text{ in.}(16.43 \text{ in.})}$$

$$= 3.5 \text{ ksi}$$

PROBLEMS

8-8. Determine the percentage error that accrues by using the average shearing stress formula instead of the exact one for the beam in Example 8-5.

8-9. Mild steel has an allowable shearing stress of approximately 14.5 ksi. Determine the average shearing stress where V is a maximum for the W30 × 173 beam in Figure 8-25. Let h = 30.44 in. and t = 0.655 in.

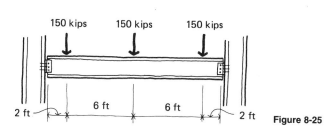

Figure 8-25

9

deflection

9-1 INTRODUCTION

The *deflection* of beams is an important topic in structural design. As noted previously, the design of a beam for a particular load generally involves the investigation of bending stresses, shearing stresses, and deflection. Building codes limit the permissible deflection of a beam just as stresses or loads are limited Some typical values are given in Table 9-1.

Excessive deflection can cause cracking of nonstructural materials that are attached to beams. Many cracks in nonbearing partition walls are due to such deflection. Doors and windows can bind up or become inoperable due to distortion of their openings by structural deflection. More important, flat or nearly flat roof surfaces are subject to "ponding," a continued buildup of water that can result eventually in a dishlike collapse.

Table 9-1 Typical Deflection Limitations Expressed as a Fraction of the Span

	TOTAL LOAD	LIVE LOAD ONLY
*Roof beams	$\dfrac{L}{180}$	$\dfrac{L}{240}$
Floor beams	$\dfrac{L}{240}$	$\dfrac{L}{360}$

*Floor-beam values should be used in place of these if a plaster ceiling is attached directly to the structural members.

Deflection is particularly critical in situations where a large portion of the total load is dead as opposed to live. Timber or concrete beams will both creep if subjected to permanent loading and will eventually sag enough to become unsightly and possibly unsafe. Therefore, when a beam supports a heavy wall or roof, for example, special design consideration needs to be given to deflection control.

Floor beams that are closely designed for bending stresses but not adequately limited in deflection can often be too ''springy'' or ''bouncy'' when loaded by impact or vibrated by a machine or vehicle to a natural frequency. This ''springiness,'' although seldom a real structural safety problem, can be most annoying and in some cases make a space unfit for occupancy.

In the last 50 years we have been able to produce steels that are very strong in tension and compression; however, the stiffness (E) of these steels remains at about 29 million psi, the stiffness of mild steel, and is independent of strength. This means that as strength increases, a larger percentage of steel beams will be designed with deflection controlling (or governing) as opposed to flexure or shear.

Beam deflections may be calculated easily by using deflection formulas available in a number of structural handbooks and design aids. A few simple cases are given in Appendix K. These formulas can also be used to approximate deflection magnitudes, for more complicated loading patterns and conditions, by ''modifying'' the actual conditions in the conservative direction so as to ''fit'' one of the tabled situations. The results so obtained will be overestimates of actual deflections and will enable the designer to ascertain if further and more accurate computations are necessary. To do this with any accuracy, an exposure to beam deflection theory is helpful, and for this reason further discussion of this technique will be deferred to a later section.

A knowledge of deflection theory will also help the designer to visualize more easily the deflected shapes of beams, frames, and other structures, determinate or indeterminate. Often a reasonably accurate image (or sketch) of how a structure deforms under load will help immeasurably in understanding how the loads are being resisted. Points of high and low stress can be ascertained, and this in turn can help decide whether a given structural choice is rational or irrational for those loads.

Deflection theory involves the study of the slopes and deflections of the neutral axis upon application of the loads. The deflected position taken by the neutral axis is called the *elastic curve*.

9-2 MOMENT-AREA METHOD

To introduce deflection theory the writer prefers the moment-area method because it emphasizes the relationship between the area under the moment curve and the resulting beam deflections and because it is often useful in providing a background for the future study of indeterminate structures.

The method is a semigraphical one first developed by Barré de Saint-Venant, a French scientist, and makes use of curves called *M/EI diagrams*. Beam deflection

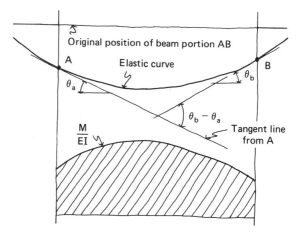

Figure 9-1   Portion of elastic curve and *M/EI* diagram.

is inversely proportional to both E and I, and an *M/EI* diagram is nothing more than a moment diagram in which every ordinate has been divided by E and I. (In this treatment it is assumed that the product EI remains constant over the length of a given beam.)

The entire method can be stated in two theorems with reference to Figure 9-1. In general the method finds slopes and deflections only *indirectly*, and careful attention should be paid to the equivalencies presented in the theorems. Proof of the theorems may be found in Appendix G.

First moment-area theorem: The change in slope between any two points, A and B, on the elastic curve is equal to the net area under the *M/EI* curve between those two points.

(Note that this theorem finds only a *change* in slope and does not directly find a slope.)

Second moment-area theorem: The vertical distance from point B on the elastic curve to a tangent line from point A is equal to the statical moment of the net area under the *M/EI* curve between points A and B taken about the vertical line through B.

(Note that this theorem finds only a vertical distance, often called a *tangential deviation*, and does not directly find a deflection.)

The theorems and the notes will both become clear after a few example problems. (Note that in these examples the symbol ∇, normally a differential operator, is used to mean "triangle side.")

Example 9-1

Determine the slope at the left end and the deflection at midspan for the beam in Figure 9-2.

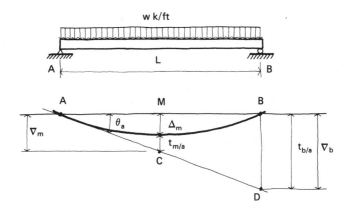

Figure 9-2 Simple beam with a uniform load.

SOLUTION: To find θ_a, it is helpful to note that $\tan \theta_a = \theta_a$ for very small angles, and that

$$\tan \theta_a = \frac{\nabla_b}{L}$$

in Figure 9-2, where ∇_b, the triangle side BD of triangle ABD, is geometrically equal to $t_{b/a}$, a tangential deviation. $t_{b/a}$ is a "vertical distance from point B on the elastic curve to a tangent line from point A," and we can use the second moment-area theorem to find it.

The M/EI curve for the beam is shown in Figure 9-3, and it is easy to compute its statical moment about a vertical line through B. See Appendix F for areas and centroidal distances of parabolic curves.

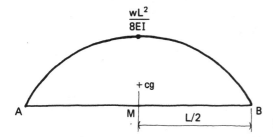

Figure 9-3 M/EI curve for the beam in Figure 9-2.

$$t_{b/a} = \frac{2}{3}\left(\frac{wL^2}{8EI}\right)(L)\frac{L}{2} = \frac{wL^4}{24EI} = \nabla_b$$

$$\tan \theta_a = \theta_a = \frac{\nabla_b}{L} = \frac{wL^4/24EI}{L}$$

$$\theta_a = \frac{wL^3}{24EI}$$

(The sign of θ_a is not evident from the computations but is negative by inspection.)

To find the midspan deflection, Δ_m (also negative by inspection), we will find the values of ∇_m and $t_{m/a}$ as shown in Figure 9-2. A subtraction will then give us Δ_m. First find ∇_m, the triangle side MC, by using θ_a.

$$\theta_a = \tan \theta_a = \frac{\nabla_m}{L/2}$$

or

$$\nabla_m = \theta_a\left(\frac{L}{2}\right) = \frac{wL^3}{24EI}\left(\frac{L}{2}\right)$$

$$= \frac{wL^4}{48EI}$$

(Note that ∇_m could also have been found by using the similar triangles AMC and ABD.)

The value $t_{m/a}$ can be found using the second moment-area theorem. From Figure 9-4,

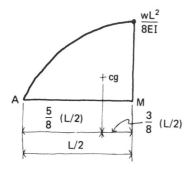

Figure 9-4 M/EI curve between A and M.

$$t_{m/a} = \frac{2}{3}\left(\frac{wL^2}{8EI}\right)\frac{L}{2}\left(\frac{3}{8}\right)\frac{L}{2}$$

$$= \frac{wL^4}{128EI}$$

Then

$$\Delta_m = \nabla_m - t_{m/a}$$

$$= \frac{wL^4}{48EI} - \frac{wL^4}{128EI}$$

$$= \frac{5wL^4}{384EI}$$

Example 9-2

Determine the slope and deflection at the free end of the cantilever beam in Figure 9-5.

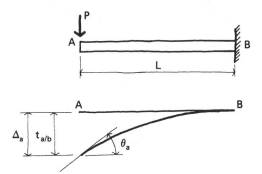

Figure 9-5   Cantilever beam with a point load.

SOLUTION: Notice that the slope of the elastic curve is zero at the fixed end. Knowing that θ_b is zero, θ_a can easily be found using the first moment-area theorem. The area under the M/EI curve between points A and B is equal to the change in θ between those same two points (Figure 9-6).

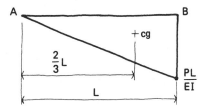

Figure 9-6 M/EI curve for the beam in Figure 9-5.

$$\Delta\theta_{ab} = \theta_a - \theta_b = \frac{1}{2}\left(\frac{PL}{EI}\right)L$$

Since $\theta_b = 0$,

$$\theta_a = \frac{PL^2}{2EI}$$

The sign is plus by inspection.

To find Δ_a, notice that a tangent line drawn from B will be coincident with the initial position of the beam before loading. This means that deflections can be found directly because they are geometrically equal to tangential deviations.

$$\Delta_a = t_{a/b} = \frac{1}{2}\left(\frac{PL}{EI}\right)L\left(\frac{2}{3}L\right)$$

$$= \frac{PL^3}{3EI}$$

The selection of the tangent line location in this example gives us a hint as to how we could have set up the simple beam of Example 9-1 so as to reduce the numerical work involved.

In Figure 9-7, we see that the slope of the beam is zero at midspan due to symmetry. With θ_m equal to zero, θ_a can be found directly as the difference between θ_m and θ_a. The midspan deflection Δ_m can also be found directly by noting that it is geometrically equal to $t_{a/m}$, which can be found by one application of the second moment-area theorem. Referring to Figure 9-4 and taking the statical moment about a vertical line through A, we get

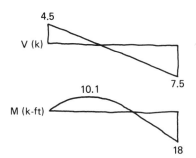

Figure 9-7

$$\Delta_m = t_{a/m} = \frac{2}{3}\left(\frac{wL^2}{8EI}\right)\frac{L}{2}\left(\frac{5}{8}\right)\frac{L}{2}$$

$$= \frac{5wL^4}{384EI}$$

Moment-area computations can often be simplified through judicious selection of tangent-line locations.

Example 9-3

Determine θ_b and Δ_m for the 4 × 16 timber beam in Figure 9-8. Let $E = 1.7(10)^6$ psi and $I = 1034$ in^4.

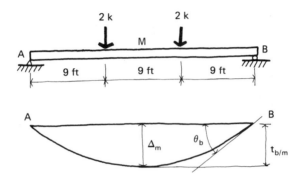

Figure 9-8 Simple beam with two concentrated loads.

SOLUTION: (It is easier to work with the symbols E and I in the computations and replace them with numerical values as a final step.)

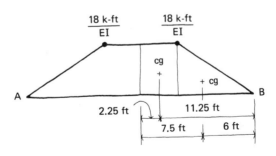

Figure 9-9 *M/EI* diagram for the beam in Figure 9-8.

$$\Delta\theta_{bm} = \theta_b - \theta_m$$

The slope at midspan is zero; therefore,

$$\theta_b = \Delta\theta_{bm}$$

$$\theta_b = \left[\frac{18 \text{ kip-ft}}{EI}(4.5 \text{ ft}) + \frac{1}{2}\left(\frac{18 \text{ kip-ft}}{EI}\right)(9 \text{ ft}) \right]$$

$$= \frac{162 \text{ kip-ft}^2}{EI}$$

The deflection at midspan is equal to $t_{b/m}$.

$$t_{b/m} = \frac{18 \text{ kip-ft}}{EI}(4.5 \text{ ft})(11.25 \text{ ft}) + \frac{1}{2}\left(\frac{18 \text{ kip-ft}}{EI}\right)(9 \text{ ft})(6 \text{ ft})$$

$$\Delta_m = \frac{1400 \text{ kip-ft}^3}{EI}$$

The problem is completed by replacing E and I by their numerical values. Notice that the units of the numerator in each case must be converted from kips and feet to pounds and inches.

$$\theta_b = \frac{162 \text{ kip-ft}^2(1000 \text{ lb/kip})(12 \text{ in./ft})^2}{1.7(10)^6 \text{ psi}(1034 \text{ in}^4)}$$

$$= 1.33$$

This is the slope of the beam at B in radians. If degrees are desired, it can be multiplied by $180/\pi$ to get

$$\theta_b = 0.76 \text{ degree}$$

To find the deflection,

$$\Delta_m = \frac{1400 \text{ kip-ft}^3(1000 \text{ lb/kip})(12 \text{ in./ft})^3}{1.7(10)^6 \text{ psi}(1034 \text{ in}^4)}$$

$$= 1.37 \text{ in.}$$

PROBLEMS

9-1. Determine the free-end slope and deflection values in terms of w, L, E, and I for a cantilever beam with a uniform load.

9-2. The beam in Figure 9-10 is a nominal 4×12 of Douglas fir. Will the deflection meet a code limitation of $L/240$?

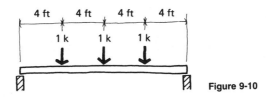

Figure 9-10

9-3 PRINCIPLE OF SUPERPOSITION

Many structures, including simple beams, are often acted upon by more than one load. It may be to advantage to treat the effects of such loads separately and add the results obtained to arrive at a final answer. This is referred to as using the *principle of superposition*. We must remember that building structures are (hopefully) never loaded such that any material reaches its yield limit or passes out of the region of elasticity. Therefore, the design loads could be placed on the structure one at a time or all at once and the resulting stresses and deflections would be the same. This idea can also be utilized from the opposite standpoint.

Suppose, for example, that we were asked to obtain the midspan deflection of the beam in Figure 9-11. Since there is no known point of zero slope, we could not obtain the answer with a single application of the second moment-area theorem. However, if we added a fictitious load of 2 kips at a point 9 ft from the right end, the loading would then be symmetrical and the problem is simply solved as in Example 9-3. The answer thus obtained would be exactly twice the true deflection that would result from the original 2-kip load.

$E = 1.7(10)^6$ psi
$I = 1034$ in.^{4}

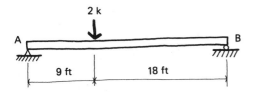

Figure 9-11 Simple beam with one concentrated load.

PROBLEMS

9-3. A simply supported W24 × 76 is 50 ft long and carries a uniform load of 2 kips/ft over the left half of its span. Determine the midspan deflection.

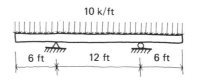

Figure 9-12

9-4. Determine the free-end deflection of the beam in Figure 9-12. Let $E = 29(10)^6$ psi and $I = 300$ in^4. (*Hint*: Use the principle of superposition to determine the deflection in parts. First find the upward deflection due to the load between the supports; then find the downward deflection due to the two overhanging loads. Algebraically add these values and then substitute for E and I.)

9-4 USE OF DEFLECTION FORMULAS

Many loading patterns and support conditions occur so frequently in construction that reference manuals and engineering handbooks tabulate the appropriate formulas for their deflections. A few such cases are given in Appendix K. More often than not, the required deflection values in a beam design situation can be obtained via these formulas, and one does not have to resort to deflection theory. Even when the actual loading situation does not match one of the tabulated cases, it is sufficiently accurate for most design situations to approximate the maximum deflection by using one or more of the formulas.

For such purposes it is helpful to know that in the case of a beam simply supported at its ends, the point of greatest deflection will always be *very* close to midspan. This is true regardless of the pattern or placement of loads along the beam. The curvature of beams is very slight, and even when a point load is placed near one end, the maximum deflection is just a little bit greater than the deflection at midspan.

Example 9-4

Determine the approximate maximum deflection for the beam in Figure 9-13.

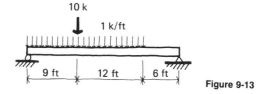

Figure 9-13

SOLUTION: Because of the relatively complex loading pattern the determination of the real value of the maximum deflection could become somewhat involved. However, we know that the midspan deflection will be very close to the maximum. Furthermore, it would be conservative in this case to treat this beam as though it had a uniform load over the *full* span and had the concentrated load located at *midspan*. The midspan deflection for this fictitious situation can easily be computed by using cases 3 and 4 in Appendix K.

$$\Delta_{max} = \frac{PL^3}{48EI} + \frac{5wL^4}{384EI}$$

or

$$\Delta_{max} = \frac{10 \text{ kips}(27 \text{ ft})^3}{48EI} + \frac{5(1 \text{ kip/ft})(27 \text{ ft})^4}{384EI}$$
$$= \frac{4100 \text{ kip-ft}^3}{EI} + \frac{6900 \text{ kip-ft}^3}{EI}$$
$$= \frac{11\,000 \text{ kip-ft}^3}{EI}$$

The value thus obtained will be slightly larger than the actual maximum deflection. If this value falls within the code limitation or is even close to it, no further deflection investigation is warranted.

Example 9-5

Determine the approximate maximum deflection for the 28-ft portion of the beam in Figure 9-14.

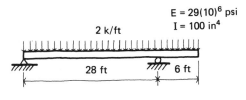

Figure 9-14

SOLUTION: In this case it would be "safe" to ignore the effect of the load on the overhang (which would produce an upward deflection on the main span) and assume a 28-ft uniformly loaded simple span.

$$\Delta_{max} = \frac{5wL^4}{384EI}$$

$$= \frac{5(2 \text{ kips/ft})(28 \text{ ft})^4(1000 \text{ lb/kip})(12 \text{ in./ft})^3}{384[29(10)^6 \text{ psi}](100 \text{ in}^4)}$$

$$= 0.96 \text{ in.}$$

Example 9-6

The W21 $\times$ 44 roof beam shown in Figure 9-15 is adequate in moment. Will it meet a deflection limitation of $L/180$?

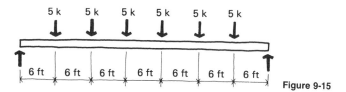

Figure 9-15

SOLUTION: The total load on the beam is 6 times 5 or 30 kips. If this load were uniformly spread out, the unit load would be $w = 0.71$ kip/ft. The deflection due to this fictitious load will be less than the actual deflection. If the loads were all gathered into one load of 30 kips and applied at midspan, the deflection thus generated would be considerably greater than the actual.

$$\Delta_{min} = \frac{5wL^4}{384EI}$$

$$= \frac{5(0.71 \text{ kip/ft})(42 \text{ ft})^4(1000 \text{ lb/kip})(12 \text{ in./ft})^3}{384[29(10)^6 \text{ psi}](843 \text{ in}^4)}$$

$$= 2.0 \text{ in.}$$

$$\Delta_{max} = \frac{PL^3}{48EI}$$

$$= \frac{30 \text{ kips}(42 \text{ ft})^3(1000 \text{ lb/kip})(12 \text{ in./ft})^3}{48[29(10)^6 \text{ psi}](843 \text{ in}^4)}$$

$$= 3.3 \text{ in.}$$

We now know that the actual beam deflection is between 2.0 and 3.3 in. and, in view of the loading pattern, is probably closer to the lesser value. The code limit is

$$\Delta_{code} = \frac{L}{180} = \frac{42 \text{ ft}(12 \text{ in./ft})}{180} = 2.8$$

Examining the upper and lower limits of the actual deflection versus the code allowable, this beam is probably OK in deflection.

PROBLEMS

9-5. A W16 × 31 serves as a simple beam 30 ft long. It supports a uniformly varying load that varies linearly from zero at one end to 1.5 kips/ft at the other. Determine the maximum deflection.

9-6. A Douglas fir 4 × 12 beam spans 20 ft and is loaded only over its central 10 feet by a uniform load of 200 p/f. Determine the approximate maximum deflection.

9-5 SUPERPOSITION AND INDETERMINATE STRUCTURES

A proper investigation of statically indeterminate structures is beyond the scope of this text. However, certain structures can be readily approached using only the ideas of superposition and the equations of statics. For example, beams that are indeterminate to the first degree (those having only one redundant support component) are easily analyzed.

Example 9-7

Determine the vertical reactions for the indeterminate beam in Figure 9-16.

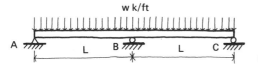

Figure 9-16 Two-span continuous beam.

SOLUTION: From statics we know that $A_y + B_y + C_y = 2wL$. Moment equations, however, cannot be used directly because any selected moment center will only eliminate one of the three forces and still leave two independent unknowns in each equation (Figure 9-17). It is noted that if any one of the three forces could be obtained by some other means, the remaining two can be easily evaluated. For example, if we denote reaction B_y as the redundant force and remove it, the beam will deflect as shown in Figure 9-18.

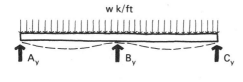

Figure 9-17 Three unknown support forces.

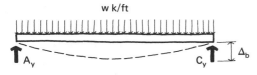

Figure 9-18 Deflected beam without B_y.

If we now apply a force P vertically upward at B, the beam will be pushed back toward its original position. Indeed, if we apply just the right amount of P, say equal in magnitude to B_y, we will then have the deflected shape as shown in Figure 9-19, having reduced Δ_b to zero. In other words, the amount of P necessary to remove the deflection Δ_b is called B_y.

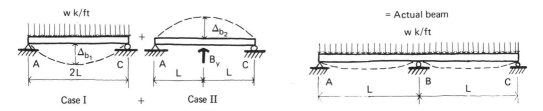

Figure 9-19 Loading cases for Example 9-7.

The procedure is to first remove the redundant force and calculate the deflection at that point, Δ_{b_1}. The deflection must be equal to Δ_{b_2}, the upward deflection due to B_y acting alone, if we are to get back to the real beam situation of zero deflection at B.

$$\Delta_{b_1} = \Delta_{b_2}$$

Using the appropriate deflection equations from Appendix K, we get

$$\frac{5w(2L)^4}{384EI} = \frac{B_y(2L)^3}{48EI}$$

E, I, and L^3 will drop out, leaving us with

$$B_y = 1.25wL$$

Through symmetry and statics we can then find that

$$A_y = C_y = 0.375wL$$

It should be noted any one of the three vertical support forces could have been declared as the redundant in this example. Owing to a lack of symmetry, the arithmetic would be a bit more involved if we had chosen A_y or C_y.

The concept of equating deflections (really superposition of loads) is a very useful tool in structural analysis. The following example will illustrate its application to a very different kind of indeterminate structure.

Example 9-8

Figure 9-20 shows two beams crossed at midspan and having the same *EI* value. Beam *A*, however, is twice as long as beam *B*. How much of the 90 kips is carried by each beam?

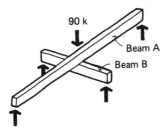

Figure 9-20 Crossed simple beams.

SOLUTION: The key to the solution is to recognize that the midspan deflection of the two beams will be equal. From statics the amount of load carried by the two beams must sum to 90 kips. If the load carried by beam *B* is called P_b, the *net* load carried by beam *A* is $90 - P_b$, as shown in Figure 9-21. Beam *A* is simultaneously acted upon by 90 kips down and the contact force P_b up, whereas beam *B* is loaded only by P_b downward.

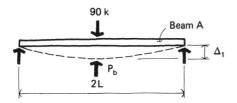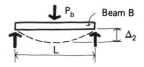

Figure 9-21 Free-body diagrams.

$$\Delta_1 = \Delta_2$$

$$\frac{(90 - P_b)(2L)^3}{48EI} = \frac{P_b(L)^3}{48EI}$$

Solving for P_b yields

$$P_b = 80 \text{ kips}$$

This indicates that the long beam is very lightly loaded, carrying only 10 kips. This is not surprising once we realize that the short beam is much stiffer; consequently it takes considerably more load to deflect than does the long beam. The support reactions will be 40 kips and 5 kips for the short and long beams, respectively.

Example 9-8 is often used to explain the behavior of a rectangular monolithic concrete slab supported on all four edges. The more rectangular or less square is the slab, the greater is the fraction of the load taken by the long edge supports that make the short span. While the bending and torsional forces in a monolithic slab are more involved than this, the idea is essentially correct.

PROBLEMS

9-7. Determine the reactions and the shear and moment diagrams for the beam in Figure 9-22. (*Hint*: *EI* need not be known.)

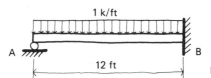

Figure 9-22   Propped cantilever.

9-8. In Figure 9-23, beams *A* and *B* are crossed at 90° in plan. The two beams are made of the same wood, but beam *A* has twice the *I* value of beam *B*. Determine how much of the load is taken by each beam.

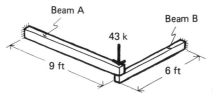

Figure 9-23   Crossed cantilevers.

9-9. A beam of constant *EI* is continuous over three walls, making two equal spans. There is a concentrated load *P* applied at the center of each span. Determine the amount of reactive force provided by each wall in terms of *P*.

10

elastic
buckling
of
columns

10-1 COLUMNS AS BUILDING STRUCTURAL ELEMENTS

Columns are probably the most important of the various structural elements in the conventional building frame. Stacked vertically, one on top of the other, they receive live and dead loads at each floor level and must transmit these loads to the foundation system below. The spacing of columns in plan usually determines what is referred to as the *structural bay*; for example, if columns are spaced 24 ft on center in one direction and 32 ft in the other, the structural bay size is said to be 24 × 32 ft. The size and shape of the structural bay has a great influence upon the type of framing system to be used in the floor structure.

Columns are usually designed with greater factors of safety than other structural elements, because any column failure would result in the catastrophic collapse of at least a major portion of the building frame. When a column fails, any beams or girders framing into it come down, as do all the other columns directly above, as shown in Figure 10-1.

Depending upon the skill of the designer in the initial stages of structural planning, columns can serve as valuable organizers or as ill-located hindrances to the architectural design process. Columns can be effective space dividers or modulators in large areas. Indeed, more columns than are needed structurally are sometimes used to separate one space from another functionally while preserving visual continuity.

It is most important that the designer or structural planner understand the structural behavior of columns under various kinds of loading. How long and slender

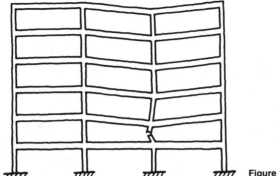

Figure 10-1 Second-story column failure.

can a given column be and still have a useful load capacity? When is a rectangular cross section more appropriate than a square one? How can the column work together with the horizontal spanning members that frame into it? What happens when the center of a column is removed so that services can be run vertically through it? How do the top and bottom connections affect column capacity? Can column capacity always be increased by using a stronger material? The structural planner should not only be able to answer these questions and others but should also have a real understanding of the principles that generate those answers.

Many codes include a provision for reducing the total design live load on certain structural members which have large tributary areas. This reduction is based on the relatively low probability that the entire area will ever be loaded to the full design value. Although this can be applicable to lower-story columns, which often support many square feet of floor area, such load reductions have been purposely ignored in the examples and problems of this chapter. It is felt that such a provision could easily lead to large errors in the nonconservative direction during the preliminary design stages.

10-2 COLUMN FAILURE MODES

Columns are essentially compression elements and, when overloaded sufficiently, will fail by crushing or buckling or a combination of these two effects (Figure 10-2). Very short stout columns will fail by crushing, and long slender columns will fail by buckling. Actually, most columns in buildings are proportioned such that both effects would be involved.

Pure crushing is a relatively easy concept to understand and a very simple design formula is available to prevent its occurrence. The designer merely provides enough cross-sectional area in the column so that the allowable compressive stress for the material is not exceeded.

$$A_r = \frac{P}{F} \tag{10-1}$$

where A_r = area required (in^2)

       P = total load on the column (kips or lb)

       F = allowable compressive stress (ksi or psi)

       The allowable bearing stress is obtained by reducing the value of the actual crushing strength for a material by an appropriate factor of safety. (It should be noted that in the case of most steel column shapes, such crushing does not occur because a similar type of failure called *local crippling* or *buckling* occurs under a lesser load. An example of such a failure occurs when we "crush" a tin can vertically without actually crushing the material.)

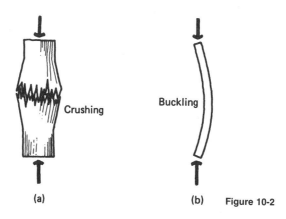

(a)   Crushing   Buckling   **(b)**   **Figure 10-2**

10-3   THE EULER THEORY

Pure buckling or elastic buckling of long, slender columns is not so easy to understand. Here, column capacity is dependent upon the dimensions and shape of the column and upon the stiffness of the material. Surprisingly enough, pure buckling is totally independent of the strength of the material!

       The basic theory of elastic buckling was successfully formulated over 200 years ago by Leonhard Euler (1707–1783), a Swiss mathematician. Essentially, such buckling occurs because there exists more than one position of equilibrium for a long, straight compression member. The slightly deflected column shown in Figure 10-3 could carry a load and be in equilibrium just like the straight one. Conversely, this could never happen in a tension member.

       As you gradually increase the axial load on a long column that is initially

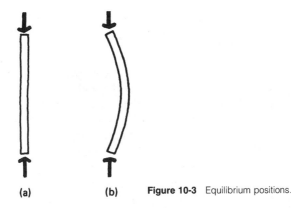

(a)   **(b)**   **Figure 10-3** Equilibrium positions.

straight, it will suddenly deflect laterally. If the load is removed, the column will return to its initial straight shape. This behavior is called *elastic buckling*. The particular value of axial load (called the critical load), which causes buckling, is given by the *Euler equation*:

$$P_{cr} = \frac{\pi^2 EI}{L^2} \tag{10-2}$$

where P_{cr} = axial load necessary to cause buckling (lb)
  E = modulus of elasticity of the column material (psi)
  I = moment of inertia of the column cross section (in^4)
  L = length of the column (in.)

[See Appendix C for the derivation of Equation (10-2).]
  Experimentally, most long columns buckle under loads that are a bit less than the value given by this equation. This is due to a number of factors, such as slight irregularities in the straightness of the member, slight eccentricity of the load, and the material being nonhomogeneous.
  This equation is more useful if we make a minor modification using the radius-of-gyration concept from Section 3-5,

$$r = \sqrt{\frac{I}{A}}$$

where r = radius of gyration of the column cross section (in.)
  A = area of the column cross section (in^2)

If we solve this expression for I and substitute into Equation (10-2),

$$\left(\frac{P}{A}\right)_{cr} = \frac{\pi^2 E}{(L/r)^2} \tag{10-3}$$

where $(P/A)_{cr}$ is the stress caused by the critical load.

The parameter L/r is called the *slenderness ratio*. The critical stress (and load) is inversely proportional to the square of this ratio.

Equations (10-2) and (10-3) are equally valid and can be used interchangeably. As stated previously, notice that neither one includes a term representing the strength of the material. Also notice that the equations give the failure loads and stresses. No factor of safety has been included.

Euler's theory assumes that the column has pinned ends that allow the column ends to rotate freely but not to translate. If the formulas are applied to columns having other end conditions, they must be adjusted by a factor to account for this change. For the following example problems of this section we shall assume pinned ends.

Example 10-1

Determine the critical buckling stress and load for a wood 4 × 4 post that is 10 ft long. Assume that $E = 1.2(10)^6$ psi.

SOLUTION: Properties of timber sections may be found in Appendix I.

$$\left(\frac{P}{A}\right)_{cr} = \frac{\pi^2 E}{(L/r)^2}$$

$$
\begin{aligned}
A &= 12.3 \text{ in}^2 \\
I &= 12.5 \text{ in}^4 \\
r &= 1.01 \text{ in. (computed as } r = \sqrt{I/A}) \\
\frac{L}{r} &= \frac{10 \text{ ft}(12 \text{ in./ft})}{1.01 \text{ in.}} \\
&= 119
\end{aligned}
$$

$$
\begin{aligned}
\left(\frac{P}{A}\right)_{cr} &= \frac{\pi^2(1.2)(10)^6 \text{ psi}}{(119)^2} \\
&= 836 \text{ psi} \\
P_{cr} &= \left(\frac{P}{A}\right)_{cr} (A) \\
&= 836 \text{ psi}(12.3 \text{ in}^2) \\
&= 10\ 300 \text{ lb}
\end{aligned}
$$

Example 10-2a

Determine the critical buckling stress for a W8 × 35 steel column that is 30 ft long.

SOLUTION: Properties of selected steel shapes may be found in Appendix J. E is taken as $29(10)^6$ psi for all rolled steel shapes. The radius of gyration values may be computed from I and A.

$$\left(\frac{P}{A}\right)_{cr} = \frac{\pi^2 E}{(L/r)^2} \qquad r_x = 3.51 \text{ in.}$$

$$r_y = 2.03 \text{ in.}$$

Compute the L/r value for each of the two axes. Substitute the larger of the two values into the Euler equation because it will yield the smaller critical stress value.

$$\frac{L}{r_x} = \frac{30 \text{ ft}(12 \text{ in./ft})}{3.51 \text{ in.}} = 103$$

$$\frac{L}{r_y} = \frac{30 \text{ ft}(12 \text{ in./ft})}{2.03 \text{ in.}} = 177$$

$$\left(\frac{P}{A}\right)_{cr} = \frac{\pi^2[29(10)^6 \text{ psi}]}{(177)^2}$$

$$= 9140 \text{ psi}$$

The use of L/r_x would clearly yield a much larger critical stress value. This indicates that the column would buckle about the y-axis (in the x direction) under a much smaller load than would be required to make it buckle the other way. In practical terms this means that, in case of overload, the column would not be able to reach the critical load necessary to make it buckle about its strong axis; it would have failed at a lower load value by buckling about its weak axis. Therefore, in computing critical load and stress values, always use the greater L/r value.

Example 10-2b

Determine the critical buckling load for the column of Example 10-2a.

SOLUTION:

$$A = 10.3 \text{ in}^2$$

$$P_{cr} = \frac{\pi^2 EI}{L^2} \qquad I_y = 42.6 \text{ in}^4$$

Since L is the same for both axes, we need only I_y for use in the equation, (i.e., L/r_y will be greater than L/r_x.)

$$P_{cr} = \frac{\pi^2[29(10)^6 \text{ psi}](42.6 \text{ in}^4)}{[(30 \text{ ft})(12 \text{ in./ft})]^2}$$

$$= 94\ 100 \text{ lb}$$

The same answer could have been obtained using the critical stress value from Example 10-2a.

$$P_{cr} = \left(\frac{P}{A}\right)_{cr} (A)$$
$$= 9140 \text{ psi}(10.3 \text{ in}^2)$$
$$= 94\ 100 \text{ lb}$$

PROBLEMS

10-1. Determine the critical buckling load for a Douglas fir 6 × 6 that is 17 ft long.
10-2. Determine the critical buckling stress for a steel pipe column that is 16 ft long. The outside diameter is 4½ in. and the wall thickness is ¼ in.
10-3. Determine the critical buckling load for a single wood 2 × 4 stud that is 8 ft long. Assume that $E = 1.3(10)^6$ psi.

10-4 INFLUENCE OF DIFFERENT END CONDITIONS

How the column ends are connected to the rest of the structure has a large influence on the critical buckling load. If the column ends are restrained from rotation in some manner, the effective buckling length can be very different from the true length, as shown in Figure 10-4. True length will be called L and the effective length KL, where K is a theoretical modifier that accounts for the effect of different end conditions.

The effective length for a column with both ends fixed is just one-half that

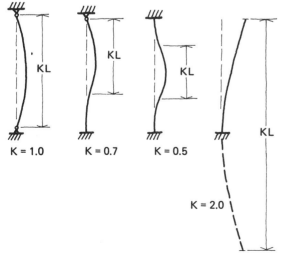

Figure 10-4 Theoretical end condition factors.

of a column with both ends pinned. The "flagpole" type of column has a K value of 2.0, which is rationalized by noting the mirror image below the fixed end needed to obtain a full buckling curve. Fortunately, this case seldom appears in building columns.

The problems presented so far have all assumed pinned ends with $K = 1.0$. Many building columns have K values that are in between the four cases shown, and judgment must be used in estimating a proper K value.

The equations presented previously should be modified to include this end condition factor.

$$P_{cr} = \frac{\pi^2 EI}{(KL)^2} \qquad (10\text{-}2a)$$

$$\left(\frac{P}{A}\right)_{cr} = \frac{\pi^2 E}{(KL/r)^2} \qquad (10\text{-}3a)$$

By noting that K is squared along with L, we can see the big difference these end conditions will make. For example, if we hold all other parameters constant and vary only K, then

1. A fixed-end column will support 4 times the load of one with pinned ends.
2. A pinned-end column will support 4 times the load of one with one fixed and one free end (flagpole).

Some of the examples that follow indicate fixed ends for timber columns. In actual construction detailing, this is quite difficult to achieve. However, such examples are included here to illustrate the use of the K factor.

Example 10-3

The timber 6 × 6 column of Figure 10-5 is 24 ft long and can be considered pinned at the lower end and effectively fixed by deep trusses framing into it at the top. Determine the critical buckling stress and load. Assume that $E = 1.5(10)^6$ psi.

SOLUTION:

$$A = 30.3 \text{ in}^2$$

$$r = 1.6 \text{ in.} \left(\text{computed as } r = \sqrt{\frac{I}{A}} \right)$$

$$K = 0.7$$

$$\frac{KL}{r} = \frac{0.7(24 \text{ ft})(12 \text{ in./ft})}{1.6 \text{ in.}}$$

$$= 126$$

$$\left(\frac{P}{A}\right)_{cr} = \frac{\pi^2 E}{(KL/r)^2}$$

$$= \frac{\pi^2[(1.5)(10)^6 \text{ psi}]}{(126)^2}$$

$$= 933 \text{ psi}$$

$$P_{cr} = \left(\frac{P}{A}\right)_{cr} (A)$$

$$= (933 \text{ psi})(30.3 \text{ in}^2)$$

$$= 28\ 300 \text{ lb}$$

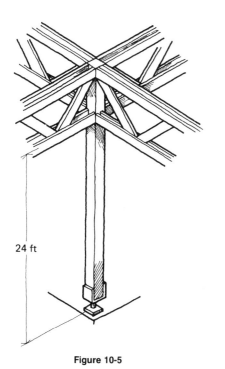

24 ft

Figure 10-5

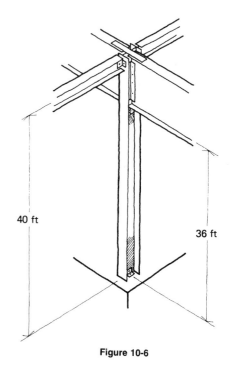

40 ft

36 ft

Figure 10-6

Example 10-4

A W8 × 67 section is used for the column in Figure 10-6. The bottom clip angle connection is a pin. Deep plate girders frame into the web, which serve to fix the weak axis at the top. Small bracing beams are clipped to the flanges and provide a pinned condition. Determine the critical buckling load.

SOLUTION: From Appendix J,

$$A = 19.7 \text{ in}^2$$
$$I_x = 272 \text{ in}^4$$
$$I_y = 88.6 \text{ in}^4$$

We can then find that

$$r_x = 3.72 \text{ in.}$$

and

$$r_y = 2.12 \text{ in.}$$

Next determine which axis is critical (i.e., which one has the greater KL/r).

$$\left(\frac{KL}{r}\right)_x = \frac{1.0(40 \text{ ft})(12 \text{ in./ft})}{3.72 \text{ in.}} = 129$$
$$\left(\frac{KL}{r}\right)_y = \frac{0.7(36 \text{ ft})(12 \text{ in./ft})}{2.12 \text{ in.}} = 143$$

The weak axis is critical for this column. We can now determine the critical buckling stress and multiply by the area to get the critical load. We can also find the load directly by substituting the critical axis properties into

$$P_{cr} = \frac{\pi^2 EI}{(KL)^2}$$
$$= \frac{\pi^2[29(10)^3 \text{ ksi}](88.6 \text{ in}^4)}{[(0.7)(36 \text{ ft})(12 \text{ in./ft})]^2}$$
$$= 277 \text{ kips}$$

Example 10-5

A steel pipe column has one end fixed and one end free. It has an outside diameter of 2.4 in. and an inside diamter of 2.0 in. It supports an axial load of 10 kips. Determine the actual length L that this column can reach without buckling.

SOLUTION:

$$I = \frac{\pi}{4}(R_o^4 - R_i^4)$$
$$= 0.843 \text{ in}^4$$
$$K = 2.0$$

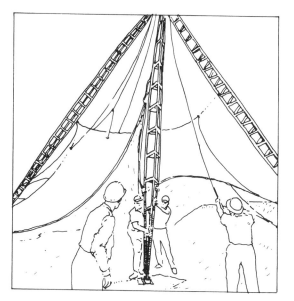

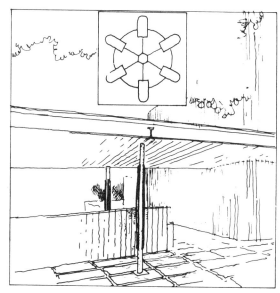

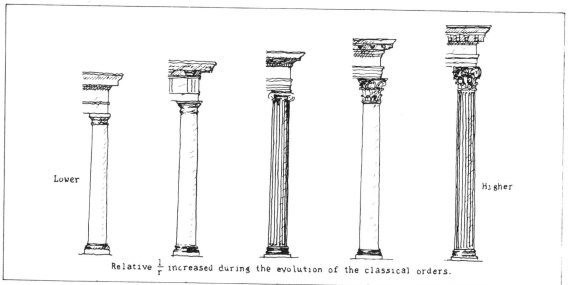

Lower   Higher

Relative $\frac{l}{r}$ increased during the evolution of the classical orders.

Eulor's formula, which correlates the factors principally responsible for a column's resistance to buckling, is another abstraction seemingly distant from "design," whose manipulation in fact has immediate visual consequences. The type of *end connections*—K in KL/r—and *column slenderness*—the L/r in the formula—can be seen as partial rationales underlying what seem to be merely arbitrary formal gestures or superficial modifications due only to changes in taste.

For example, a compression strut in Frei Otto's pavilion for the Museum of Modern Art garden displays a quite conscious gradation in cross section, center to ends, reflecting the fact that with pinned ends the middle of a column must resist the tendency to buckle. Alvar Aalto may have had a similar structural logic less directly in mind when he designed the columns in his 1937 Finnish Pavilion in Paris. Each has six ribs, tapering center to ends, added to a cylindrical section; yet the end connections are certainly not pins, and the ribs may in fact add only marginally to the columns' buckling resistance. But Aalto was using our intuitive visual knowledge of behavior under load in order to involve us with the building; the ribs are a kind of plausible structural fairy tale, an invitation to empathy.

The evolution of the classical orders can be interpreted as another instance where patterns of structure coincide with patterns of visual sophistication. Despite numerous individual exceptions, the clear pattern is one of regular increase in the slenderness ratio, from the earlier Tuscan and Doric to the later Composite. To put it another way, the increase in L/r reflects both a greater technical confidence, the result of accumulated experience with columns and loads, and a greater affinity for visual lightness produced by changes in both proportion and ornamentation.

L is defined as the length at which P becomes critical.

$$P_{cr} = \frac{\pi^2 EI}{(KL)^2}$$

or

$$L = \frac{\sqrt{\pi^2 EI/P_{cr}}}{K}$$

$$= \frac{\sqrt{\dfrac{\pi^2[29(10)^3 \text{ ksi}](0.843 \text{ in}^4)}{10 \text{ kips}}}}{2.0}$$

$$= 78 \text{ in. or } 6.5 \text{ ft}$$

PROBLEMS

10-4. A W12 × 50 shape is used as a column 50 ft long. If both ends are fixed, determine the critical buckling stress and load.

10-5. Figure 10-7 shows a 4 × 2 in. structural tube with a ¼-in. wall thickness serving as a column 16 ft long. Its upper end has pinned connections. The lower end is braced by a masonry wall so that its weak axis is fixed and the strong axis pinned. Determine the critical buckling load.

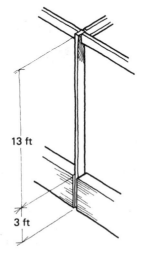

13 ft

3 ft

Figure 10-7

10-6. A 6-in.-diameter (actual dimension) wood post is fixed into a large foundation pier at grade and is completely free at its upper end. How long can it be and still just support a load of 2 kips without failing? Assume that $E = 1.3(10)^6$ psi.

10-5 INTERMEDIATE LATERAL BRACING

We now know that if the end conditions are the same for both axes, a column will always buckle about its weak axis. A rectangular timber post will buckle in a direction parallel to its least dimension. A steel wide-flange shape will buckle in a direction parallel to its flanges. With any asymmetrical shape, we have a situation in which the full capacity of the strong axis is not normally utilized. However, there are many situations in structural frames where we can increase the capacity of such asymmetrical shapes by decreasing the effective weak-axis length. In Example 10-4, this occurred to a certain degree by virtue of the different end connections. We will have a structurally more efficient column if $(KL/r)_x$ and $(KL/r)_y$ have values that are similar in magnitude. Intermediate lateral bracing members are a most effective way of doing this. Often such elements occur rather naturally for other construction reasons.

In Figure 10-8, the column is braced against weak-axis buckling by a secondary wall element. Bracing can be provided by load-carrying beams and girders as well.

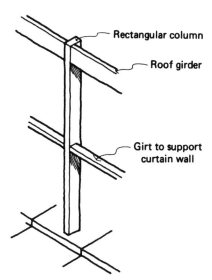

Rectangular column

Roof girder

Girt to support curtain wall

Figure 10-8 Intermediate bracing.

It is important to realize that such members do not provide any bracing for the other axis of the column. In terms of the support provided for the column, they can be considered as two-force members (i.e., unable to resist nonaxial forces).

Example 10-6

A southern pine 4 × 8 section is used as a column 20 ft long. It has pinned ends and is braced against weak-axis buckling at midheight (Figure 10-9). Determine the critical buckling stress.

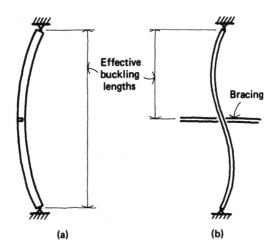

Figure 10-9 (a) Strong-axis buckling; (b) Weak-axis buckling.

SOLUTION: From Appendix H, $E = 1.6(10)^6$ psi and from Appendix I, the area and I values can be found to determine the r values.

$$r_x = 2.1 \text{ in.}$$
$$r_y = 1.01 \text{ in.}$$

$$\left(\frac{KL}{r}\right)_x = \frac{1.0(20 \text{ ft})(12 \text{ in./ft})}{2.1 \text{ in.}} = 114$$
$$\left(\frac{KL}{r}\right)_y = \frac{1.0(10 \text{ ft})(12 \text{ in./ft})}{1.01 \text{ in.}} = 119$$

The weak axis is critical, but not by much.

$$\left(\frac{P}{A}\right)_{cr} = \frac{\pi^2 E}{(KL/r)^2}$$
$$= \frac{\pi^2[(1.6)(10)^6 \text{ psi}]}{(119)^2}$$
$$= 1120 \text{ psi}$$

Example 10-7

A W10 × 49 is used as a 40-ft-long column. It has pinned ends and its weak axis is braced at a point 22 ft up from the lower end (Figure 10-10). Determine the critical buckling stress and load.

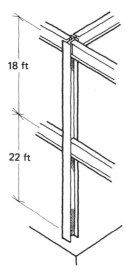

Figure 10-10

SOLUTION: Values from Appendix J are

$$I_x = 272 \text{ in}^4$$
$$I_y = 93.4 \text{ in}^4$$
$$A = 14.4 \text{ in}^2$$

We can then find that

$$r_x = 4.35 \text{ in.}$$

and

$$r_y = 2.54 \text{ in.}$$

Next determine the larger KL/r.

$$\left(\frac{KL}{r}\right)_x = \frac{1.0(40 \text{ ft})(12 \text{in./ft})}{4.35 \text{ in.}} = 110$$
$$\left(\frac{KL}{r}\right)_y = \frac{1.0(22 \text{ ft})(12 \text{ in./ft})}{2.54 \text{ in.}} = 104$$

The strong axis is critical.

$$\left(\frac{P}{A}\right)_{cr} = \frac{\pi^2 E}{(KL/r)^2}$$
$$= \frac{\pi^2[29(10)^3 \text{ ksi}]}{(110)^2}$$
$$= 23.6 \text{ ksi}$$

$$P_{cr} = \left(\frac{P}{A}\right)_{cr} (A)$$
$$= 23.6 \text{ ksi}(14.4 \text{ in}^2)$$
$$= 340 \text{ kips}$$

PROBLEMS

10-7. A W8 × 10 section is used as a 30-ft-long column. The upper end is pinned, the lower end fixed and the weak axis is braced at midheight by two angles as shown in Figure 10-11. Determine the critical buckling stress.

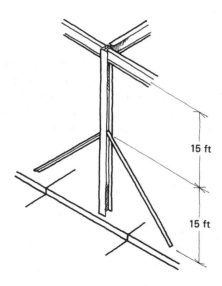

15 ft

15 ft

Figure 10-11   Diagonally braced column.

10-8. Determine the critical buckling stress and load for a wood 2 × 6 if it is 20 ft long, has pinned ends, and has its weak axis braced at 5-ft intervals. Assume that $E = 1.2(10)^6$ psi.

10-9. Figure 10-12 shows a C10 × 30 channel used as a long pinned-end compression member of length L. Determine the optimum spacing XL of intermediate bracing elements such that the critical buckling load will be the same for both axes. X will be a fraction of L.

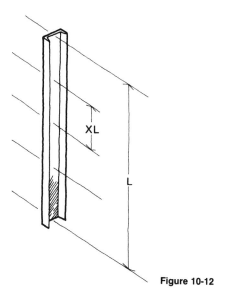

Figure 10-12

10-6 LIMITS TO THE APPLICABILITY OF THE EULER EQUATION

The plot of the Euler equation in Figure 10-13 shows that it is asymptotic to both axes. We can see that for very low values of KL/r (i.e., for short stout columns), the critical stress becomes very high. Indeed, as KL/r approaches zero, $(P/A)_{cr}$ goes to infinity. Obviously, this cannot be valid because the stresses in this region of the graph would be above the yield stress or crushing stress of the material. It is clear

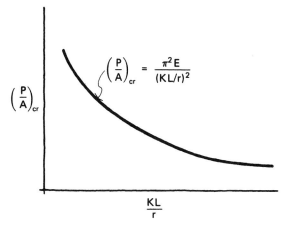

Figure 10-13 Euler equation.

that the Euler equation cannot be valid if it predicts a buckling stress above the yield stress. A short column will fail by crushing under a load which is less than that predicted by the Euler equation.

The value $(P/A)_{cr}$ represents an average unit stress over the entire cross section. When a column is buckled into an arc, the stresses are not uniform over the cross section. The maximum fiber stress will be compressive, occurring on the inside of the arc. For some columns of relatively low slenderness, $(P/A)_{cr}$ may be large enough to cause a yield or crushing of these fibers. This implies that the Euler equation for elastic buckling is not completely valid except for long, thin columns. A column with a moderately (often called "intermediate") low slenderness ratio will still basically buckle rather than crush, but as this occurs some of the fibers will reach their yield stress. This type of buckling is called *inelastic buckling* and is really a combination of buckling and crushing.

In terms of developing design equations for safe loads on columns, both the timber and steel industries have come up with appropriate lower limits on KL/r. Below these values, the Euler equation cannot be used and empirically developed equations to handle these shorter columns are employed. Since this chapter is intended to treat only elastic buckling, most of the examples and problems provided involve relatively long and slender columns.

11

trusses

11-1 INTRODUCTION

A truss is a lightweight frame generally used for relatively long spans in buildings and bridges. They are usually placed parallel to one another to make a one-way system for a floor or roof deck. Their lightness means they are deeper than beams would be if used on a similar span, and for this reason trusses are more frequently used in roof structures.

In the United States, trusses are almost always constructed of wood or steel, but in other countries they have also been precast in reinforced concrete. The light triangular wood truss, made of nominal 2 × 4 and 2 × 6 elements and placed 2 ft on centers, is used almost exclusively to make residential gable roofs in some parts of the United States. It erects rapidly and enables the floor below to be free of interior bearing partitions. Steel trusses, both flat and curved, are used to span the large majority of very long span buildings such as field houses and sports domes. In such structures, self-weight can easily become a controlling design factor, and the small span/depth ratio of a truss (with its increased building envelope) becomes a welcome trade-off to minimize this dead load.

Trusses can be fabricated in almost any shape. In technical terms, a truss is a triangulated planar framework made up of linear elements that connect at pin joints. When actually constructed, these joints are seldom truly pinned, but the initial structural analysis makes this assumption anyway. (For many trusses, the members are thin and have relatively little bending resistance, so the pinned-joint assumption causes no great error.) A few of the more commonly used truss shapes are illustrated in Figure 11-1. Some have been named for the engineer or designer who popularized that particular type.

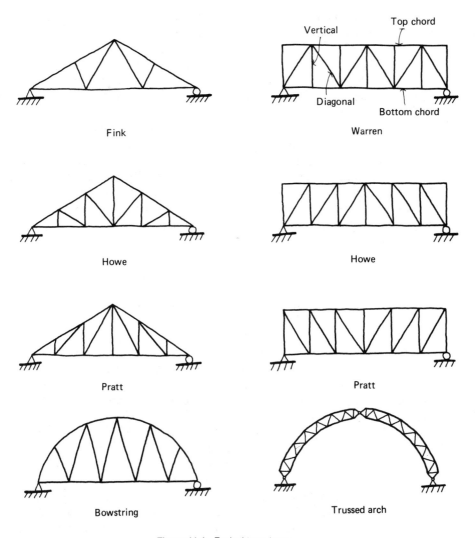

Figure 11-1 Typical truss types.

Loaded properly, each member of a truss is (in ideal terms) a two-force member. It is either in tension or compression, and if in compression, it behaves as a slender column and must be designed with elastic buckling in mind. When trusses are used on simple spans, all the top chord members form a continuous line of compression and the entire top of the truss is subject to the lateral buckling phenomenon discussed in Section 7-5. Usually, the roof "skin" provides the required lateral bracing unless the trusses are exposed. Overhanging trusses will have compression in some bottom chord elements, and these are subject to the same buckling effects.

For the purposes of preliminary design, it is assumed that trusses are loaded by concentrated loads that act only at the joints. In actuality, most floor and roof loads are uniform, and when the deck surface is attached directly to the top chord elements, these members are subjected to the combined action of axial and bending forces. The examples and problems of this chapter are concerned only with the analysis of trusses loaded through the joints, or "panel points," and all the members will be considered to carry only axial loads.

Almost all trusses are statically determinate with respect to the external reaction components. Depending on the manner of triangulation, trusses can also be determinate or indeterminate with respect to the internal forces in the members. Trusses with redundant members are internally statically indeterminate, and the member forces cannot be resolved using statics alone. Whether or not a truss is internally determinate can be ascertained by Equation (11-1). Trusses without enough members to make triangles using every joint will be unstable, and those with excess members are indeterminate.

$$m + 3 = 2j \tag{11-1}$$

where m = number of members, assuming no member runs through a joint
      j = number of joints

In Equation (11-1), the constant 3 represents the usual three external reaction components. The concept here is that the number of unknowns equals the unknown member forces (m) plus the reaction components. At a planar joint, only two force equations of equilibrium, $\sum F_x = 0$ and $\sum F_y = 0$, can be written, and this means that the total number of available equations is twice the number of joints (j).

The trusses in Figure 11-1 are determinate, as are all of the trusses in the examples and problems. Figure 11-2 shows two indeterminate trusses. The one in Figure 11-2(b) has two diagonals, which cross without a joint. This type of truss becomes determinate if we assume that those two members are so slender and flexible as to be worthless in compression, in which case only one of them will be functional, depending upon the loading pattern. The diagonals are then called *counters*.

The geometry imposed by triangulation means that, under certain loading conditions, some of the members of a truss may have no internal force. In such cases,

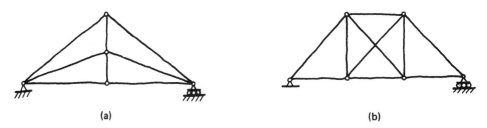

(a)   (b)

Figure 11-2 Indeterminate trusses.

the member acts as a bracing element and is usually needed for stability. These *zero members* could also carry force under a different loading pattern.

11-2 ANALYSIS BY JOINT EQUILIBRIUM

If we assume that all joints are pinned and loads and reactions act only at the joints, each joint becomes a small concurrent force system. It must be held in equilibrium by the known forces acting on it from the loads (including reactions) and by the unknown forces from the two-force members. Each joint can then be analyzed like the simple structures of Section 2-4. As pointed out previously, there are only two equations available for each joint, so we must move from joint to joint over the truss in such a manner as to be always working with only two unknowns. In many cases, this means starting at one of the joints at the ends of the truss and progressing toward the center.

The external reaction components should be determined before isolating the joints, and this has been done in the examples that follow. After solving for the reactions of a given truss, the reader should attempt to guess which members are in tension and compression before continuing with the solution. The answers obtained from any numerical analysis can then be rationalized with the visual analysis, and arithmetical errors can often be caught before they accumulate.

Example 11-1

Determine the forces in each of the members of the truss in Figure 11-3.

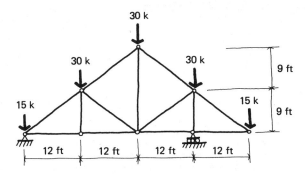

Figure 11-3 Overhanging truss.

SOLUTION: Make a free-body diagram of each joint in turn, showing compressive arrows acting toward the joint and tensile arrows pointing away from the joint. As usual, incorrect sense assumptions will result in negative answers.

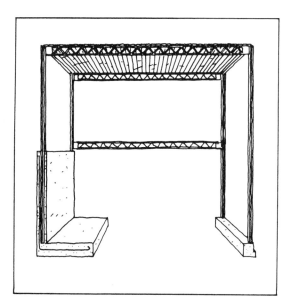

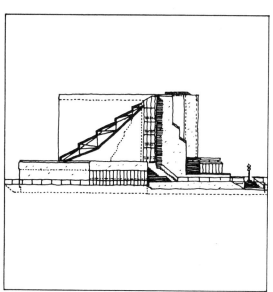

Be they straightforward or complex, trusses are relatively long-span building elements that offer considerable possibility for architectural effect along with the efficient performance of structural jobs. The variety of truss shapes, types, and adaptations is large, but even the simplest can have both visual and intellectual fascination: the attractions of a well-made puzzle, the paradox of a large strong thing made from many small weak things. Further, the usefulness of trusses in roof structures is an eternal invitation to experiments with the interacting effects of structure and light on interior space. From Charles Eames' unassumingly elegant use of stock steel bar joists in his own California house, to Frank Lloyd Wright's more spectacular (but really no more complicated) wooden trusses over the drafting room at Taliesin East, the range of expression available with simple trusses is enormous. But complexity, too, has its places and desirable effects; the technically complex trusses and the soaring space shadowed by them, in James Stirling's Cambridge History Faculty, and the decoratively complex trusses and brooding interior of Bernard Maybeck's First Church of Christ Scientist, are two diverse instances.

222

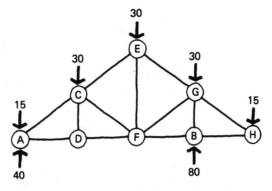

Figure 11-4 Free-body diagram of the truss.

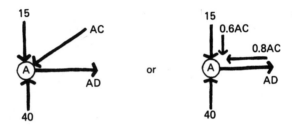

Figure 11-4(a)

$$\sum F_y = 0$$
$$40 - 15 - 0.6AC$$
$$AC = 41.7 \qquad AC = 41.7 \text{ kips C}$$

$$\sum F_x = 0$$
$$AD - 0.8(41.7) = 0$$
$$AD = 33.3 \qquad AD = 33.3 \text{ kips T}$$

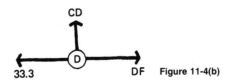

Figure 11-4(b)

$$\sum F_x = 0$$
$$DF - 33.3 = 0$$
$$DF = 33.3 \qquad DF = 33.3 \text{ kips T}$$

$$\sum F_y = 0 \qquad CD = 0$$

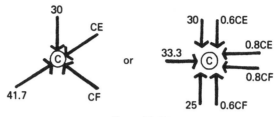

Figure 11-4(c)

$$\sum F_x = 0$$
$$33.3 - 0.8CE - 0.8CF = 0$$
$$\sum F_y = 0$$
$$-30 + 25 - 0.6CE + 0.6CF = 0$$

Solving simultaneously gives us

$$CF = 25 \qquad CF = 25 \text{ kips C}$$
$$CE = 16.7 \qquad CE = 16.7 \text{ kips C}$$

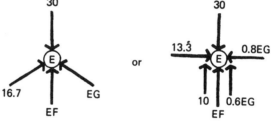

Figure 11-4(d)

$$\sum F_x = 0$$
$$13.3 - 0.8EG = 0$$
$$EG = 16.7 \qquad\qquad EG = 16.7 \text{ kips C}$$
$$\sum F_y = 0$$
$$-30 + 10 + EF + 0.6(16.7) = 0$$
$$EF = 10 \qquad\qquad EF = 10 \text{ kips C}$$

Figure 11-4(e)

$$\sum F_y = 0$$
$$-15 - 10 + 0.6FG = 0$$
$$FG = 41.7 \qquad\qquad\qquad FG = 41.7 \text{ kips T}$$

$$\sum F_x = 0$$
$$-33.3 + 20 + 0.8(41.7) + FB = 0$$
$$FB = -20 \qquad\qquad\qquad FB = 20 \text{ kips C}$$

Sense of *FB* assumed incorrectly.

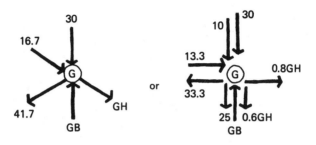

Figure 11-4(f)

$$\sum F_x = 0$$
$$-33.3 + 13.3 + 0.8GH = 0$$
$$GH = 25 \qquad\qquad\qquad GH = 25 \text{ kips T}$$

$$\sum F_y = 0$$
$$-10 - 30 - 25 - 0.6(25) + GB = 0$$
$$GB = 80 \qquad\qquad\qquad GB = 80 \text{ kips C}$$

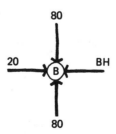

Figure 11-4(g)

$$\sum F_x = 0$$
$$-BH + 20 = 0$$
$$BH = 20 \qquad\qquad BH = 20 \text{ kips C}$$

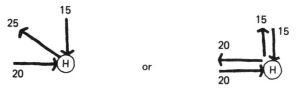

Figure 11-4(h)

Joint *H* is isolated as a check.

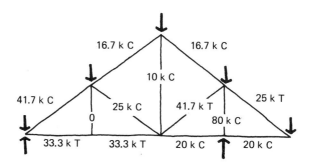

Figure 11-5 Member forces.

Example 11-2

Determine the forces in each member of the wind bent shown in Figure 11-6.

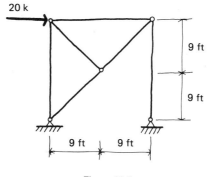

Figure 11-6

SOLUTION:

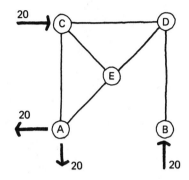

Figure 11-7 Free-body diagram

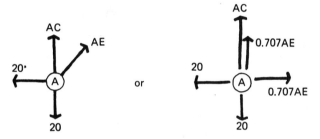

Figure 11-7(a)

$$\sum F_x = 0$$
$$-20 + 0.707AE = 0$$
$$AE = 28.3 \qquad\qquad AE = 28.3 \text{ kips T}$$

$$\sum F_y = 0$$
$$AC + 0.707(28.3) - 20 = 0$$
$$AC = 0 \qquad\qquad AC = 0$$

Figure 11-7(b)

$$\sum F_y = 0$$
$$0.707CE = 0$$
$$CE = 0 \qquad\qquad CE = 0$$

$$\sum F_x = 0$$
$$+20 - CD - 0.707(0) = 0$$
$$CD = 20 \qquad\qquad CD = 20 \text{ kips C}$$

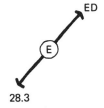

Figure 11-7(c)

$$\sum F_{AD} = 0$$
$$ED = 28.3 \qquad\qquad ED = 28.3 \text{ kips T}$$

    or

Figure 11-7(d)

$$\sum F_y = 0$$
$$DB - 20 = 0$$
$$DB = 20 \qquad\qquad DB = 20 \text{ kips C}$$

Figure 11-7(e)

Joint B is isolated as a check.

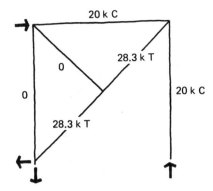

Figure 11-8 Member forces.

PROBLEMS

11-1. Determine the forces in the members of the truss in Figure 11-9.

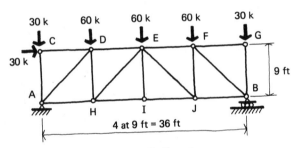

Figure 11-9 Flat Howe truss.

11-2. Determine the forces in the members of the truss in Figure 11-10.

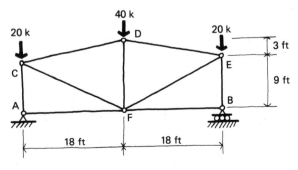

Figure 11-10

11-3.   Determine the forces in the members of the truss in Figure 11-11.

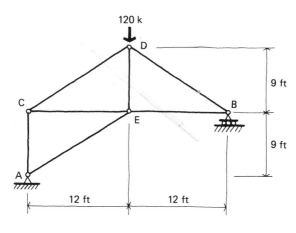

Figure 11-11

11-4.   Work Problem 11-3 after adding a 20-kip load acting to the right at joint C in Figure 11-11.

11-5.   Determine the forces in the members of the frame in Figure 11-12.

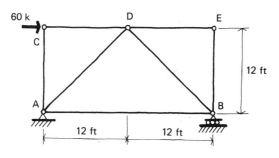

Figure 11-12

11-3   METHOD OF SECTIONS

A second approach to truss analysis, called the *method of sections*, consists of cutting sections through the truss so that a free-body diagram of a portion of the truss will involve the desired unknown member forces. In general, it is faster than the method of joints because it makes use of all three equations of planar static equilibrium. The method of joints deals with concurrent forces and therefore no moment arms are available.

Realizing that the entire truss is in equilibrium and that each joint is in equilibrium, it follows that larger portions of the truss will also be in equilibrium. If a truss is cut through by an imaginary cutting plane and a portion to one side of that plane is isolated, it will be held in a state of balance by the external forces acting on the truss and the unknown forces in the cut members. Since the free-body diagram makes these internal forces external, the equations of statics can be used to find them. With three equations available, three unknown member forces can be determined with each free-body cut.

Successive cutting planes may be used to isolate increasingly larger portions of the truss, as shown in Figure 11-13. However, one of the advantages to this method is that all the member forces need not be found if we are interested only in those in one area of the truss. (Whenever a section cuts through two concurrent unknowns, as in Figure 11-13(b), this method reduces to a joint equilibrium problem. The two procedures, of course, can be used to work different parts of the same truss.)

The senses of the forces in the unknown members are assumed so that an arrow acting against the cutting plane is compressive and one pulling away from it is tensile. As usual, a negative sign in the answer will indicate an incorrect assumption.

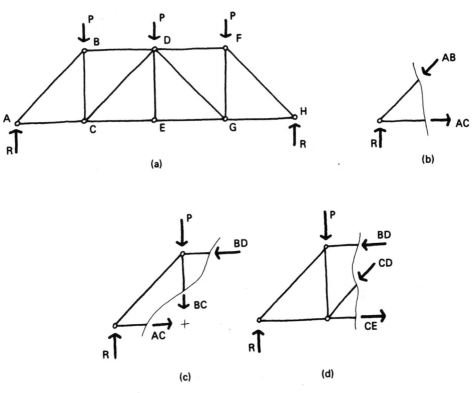

Figure 11-13 Cutting planes.

The isolated portion of the truss is treated as a rigid free body, and moment centers may be located on or off the body. The forces in cut sloping members are usually broken down into their rectangular components so that moment arms can be obtained from known dimensions. It is helpful to realize that a resultant force can be translated forward or backward along its line of action (before being replaced by its components), and this can sometimes result in convenient moment-arm distances. This translation does not in any way affect the state of equilibrium.

Example 11-3

Determine the forces in members *CE*, *DE*, and *DF* of the truss shown in Figure 11-14.

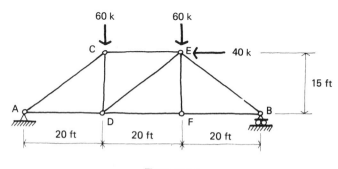

Figure 11-14

SOLUTION: First find the external reactions and then cut a section through the three members.

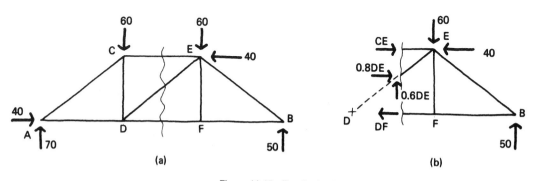

Figure 11-15   Free-body diagrams.

Applying the equations of statics to the body in Figure 11-15(b), we get

$$\sum F_y = 0$$
$$-60 + 50 + 0.6DE = 0$$
$$DE = 16.7 \qquad\qquad\qquad DE = 16.7 \text{ kips C}$$

$$\sum M_E = 0$$
$$50(20) - DF(15) = 0$$
$$DF = 66.7 \qquad\qquad\qquad DF = 66.7 \text{ kips T}$$

$$\sum M_D = 0$$
$$-CE(15) - 60(20) + 40(15) + 50(40) = 0$$
$$CE = 93.3 \qquad\qquad\qquad CE = 93.3 \text{ kips C}$$

$$\sum F_x = 0 \quad \text{(check)}$$
$$93.3 + 0.8(16.7) - 66.7 - 40 \approx 0 \quad \checkmark$$

Example 11-4

Determine the forces in members *EG*, *FG*, and *FB* of the overhanging truss in Figure 11-16.

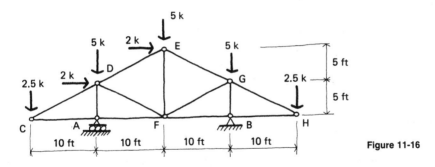

Figure 11-16

SOLUTION: Using the isolated portion of Figure 11-17(b), we get

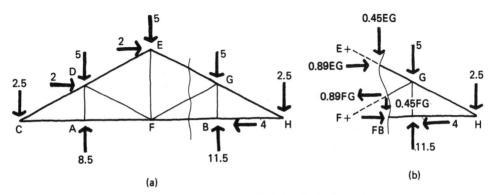

(a)   (b)

Figure 11-17 Free-body diagrams.

$$\sum M_G = 0$$
$$-4(5) - 2.5(10) + FB(5) = 0$$
$$FB = 9 \qquad FB = 9 \text{ kips C}$$

$$\sum M_F = 0$$

(Let the force EG be translated back to point E, where its vertical component will have no moment arm with respect to moment center F.)

$$-5(10) + 11.5(10) - 2.5(20) - 0.89EG(10) = 0$$
$$EG = 1.7 \qquad\qquad\qquad\qquad EG = 1.7 \text{ kips C}$$
$$\sum M_H = 0$$

(Let the force FG be translated forward to point F.)

$$5(10) - 11.5(10) + 0.45FG(20) = 0$$
$$FG = 7.2 \qquad\qquad\qquad\qquad FG = 7.2 \text{ kips T}$$

Since neither force equation was used, either will be valid for a check.

$$\sum F_x = 0 \quad \text{(check)}$$
$$0.89(1.7) - 0.89(7.2) - 4 + 9 \approx 0 \quad \checkmark$$

PROBLEMS

11-6. Use the method of sections combined with joint equilibrium to determine the forces in the members of the Warren truss in Figure 11-18.

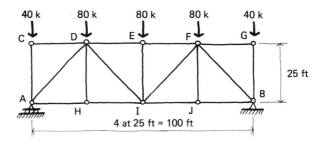

Figure 11-18

11-7. Use a cut section to determine the force in member *DE* of the truss in Figure 11-11.

11-8. Find the magnitude of the force in the tensile tie *AB* of the truss in Figure 11-19.

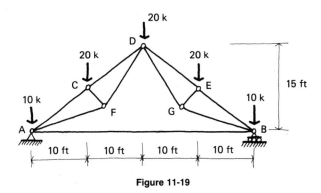

Figure 11-19

11-9. Use a cut section to find the force in member *DB* of the truss in Figure 11-12.

11-10. Determine the force in each member of the three-story wind bent of Figure 11-20.

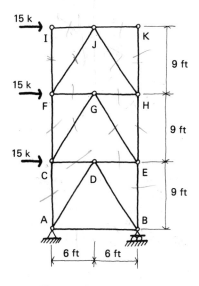

Figure 11-20   Wind bent.

11-4   SPECIAL TYPES OF TRUSSES

There are two categories of trusses which are used so frequently in construction today that they warrant special attention. One is the prefabricated light timber truss mentioned in the first section of this chapter. They are usually made up of pieces small in cross section and fastened at the joints so that all members can lie in the same plane. These joints are nailed together using special toothed light-gage-steel plates as seen in Figure 11-21. Whenever possible, chord members continue through joints for ease of fabrication. (The basic analysis still assumes a completely hinged joint, however.)

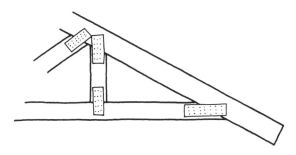

Figure 11-21 Nailing plates.

When fully braced by a roof plane, these trusses are very stiff and much stronger than they appear. Specifications have been set in a standard manner so that, if no special loads or unusual support conditions are present, these trusses can be ordered almost as stock items with obvious economy.

The second type of frequently used truss is the open-web steel bar joist, often referred to simply as a bar joist. They are usually fabricated from angles and round bar stock, with the larger ones using only angles. Shop-welded together, using continuous chord members, they are manufactured to meet certain load capacities as specified by the Steel Joist Institute. Without special detailing, they are not suitable for concentrated loads but can handle the uniform loads of most floors and flat (or nearly flat) roofs over a very wide range of spans. They are seldom designed in the usual sense. Like rolled steel beams, they are selected from load tables that have been developed with due consideration for the moments, shears, and deflections involved in simple spans.

Unlike wood joists and precast concrete joists usually used on shorter spans, open web joists easily permit the through passage of wires and pipes and even small ducts. The minimum of on-site labor required for their erection means that steel joists can often provide a very economical deck system. They are particularly well suited for one-story structures with high ceilings, such as factories and gymnasiums, where fireproofing and acoustics needs are minimal.

A third type of special truss is not really a truss at all. As shown in Figure 11-22, the Vierendeel "truss" is really more like a rigid frame or a beam with large holes. The absence of triangulation and the presence of fully moment resistant joints mean that this structure is grossly misnomered when called a truss. The Vierendeel frame takes its name from its designer, Arthur Vierendeel (1852–c.1930), a Belgian engineer and builder. It is usually made of reinforced concrete, which inherently provides the required joint fixity but can also be fabricated from steel. It carries its load through the development of bending stresses in all or most of the segments. As the frame bends, the members assume "S" patterns for their deflected shapes because the joints apply moments to the ends of the members as they rotate. This joint continuity makes the structure highly indeterminate.

While the Vierendeel frame is quite inefficient compared to a truss, it can be very useful in specific structural situations. The lack of diagonal members means that there are large clear openings in the frame that can be used functionally. Their best application occurs when the span and loads are such that a frame equal in depth to one story height is required. If the floor plan can accommodate the web verticals, the frame will become an integrated part of the architectural section. The Vierendeel frame of Figure 11-22 is used to provide a large column-free area beneath a heavily loaded second floor.

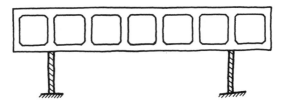

Figure 11-22 Vierendeel frame with bearing wall supports.

appendices

DERIVATION OF BASIC
FLEXURAL STRESS EQUATION

The beam in Figure A-1 has a rectangular cross section, which has been specified here for the sake of convenience and clarity. The beam cross section can be any shape and the following derivation will still be valid. It is necessary, however, to make a number of other assumptions concerning the material and the geometry, and these will be listed at the end of this appendix. The most important of these is that "planes before bending remain plane after bending," which is stated graphically in Figure A-1(b). To understand this, visualize a straight unloaded beam and make a series of imaginary slices through it quite close together. The planes made by the imaginary slicer should be parallel to each other and normal to the beam axis. When the beam is bent under load, as in Figure A-1(a), the planes will not warp or twist out of shape but will merely tilt toward one another. They will remain flat, getting closer together where the fibers are in compression and farther apart where tension occurs. This assumption is quite valid and can be proven visually using a material of low stiffness.

There will be a horizontal plane, designated *ab* in Figure A-1(b) and called the neutral axis, which will neither lengthen nor shorten. The fact that "planes remain plane" ensures that the unit strain, ϵ, will be proportional to its distance from the

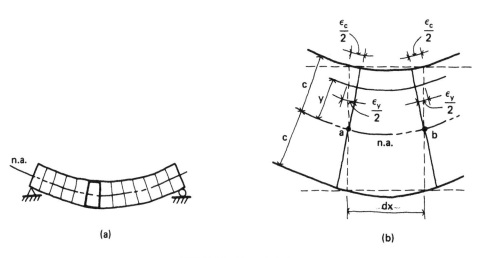

(a)

(b)

Figure A-1 Flexural strain.

neutral axis. The fiber located y distance above the n.a. in Figure A-1(b) has an original unit length of dx and has a total strain of ϵ_y. Likewise, the top fiber of the beam, c distance from the n.a., has the same original unit length and a total strain of ϵ_c. By similar triangles

$$\frac{\epsilon_y}{\epsilon_c} = \frac{y}{c} \tag{A-1}$$

Since stress is proportional to strain by Hooke's law (if we keep the stresses in the elastic region for the material), then

$$\frac{f_y}{f_c} = \frac{y}{c} \tag{A-2}$$

as illustrated in Figure A-2(c) by the triangular, straightline stress distribution.

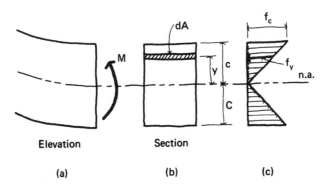

Elevation   Section

(a)   (b)   (c)

Figure A-2 Flexural stress.

The value of f_y varies from zero at the neutral axis to a maximum value of f_c at the extreme fibers of the section. Each stress f_y acts on a small area dA of beam cross section and causes a small moment about the neutral axis.

$$dM = [f_y(dA)]y$$

The summation of these small moments must equal the couple M, which is caused by the external loads. (The value of M, of course, varies along the length of a beam as represented by a moment diagram.)

$$M = \int_0^A f_y \, y \, dA$$

The variable f_y can be eliminated by using Equation A-2,

$$f_y = f_c \left(\frac{y}{c} \right)$$

giving us

$$M = \int_0^A \frac{f_c}{c} y^2 \, dA$$

Removing the constants, we get

$$M = \frac{f_c}{c} \int_0^A y^2 \, dA$$

in which the quantity $\int_0^A y^2 \, dA$ is the moment of inertia of the cross section taken with respect to the neutral axis. Making the change, we get

$$M = \frac{f_c}{c} I_{\text{n.a.}}$$

Solving for bending stress, we get

$$f_c = \frac{Mc}{I_{\text{n.a.}}} \qquad (A\text{-}3)$$

where f_c is the extreme fiber bending stress. (*Note*: f_c is often written as f_b with b understood to mean "extreme fiber bending.")

It will be proven that the neutral axis is coincident with the centroidal axis, as implied by the dimensions labeled c in Figure A-1(b). Therefore, $I_{\text{n.a.}} = I_{\text{c.g.}}$ and the subscripts are usually deleted,

$$f = \frac{Mc}{I} \qquad (7\text{-}1a)$$

The stress at some fiber other than the top or bottom locations is found as

$$f_y = \frac{My}{I} \qquad (7\text{-}1)$$

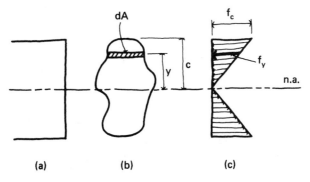

Figure A-3

To show that the neutral axis is a centroidal one, examine Figure A-3. The forces, $f_y\ dA$, must algebraically add to zero to ensure horizontal equilibrium ($\sum F_x = 0$) for the beam section on which they act.

$$\int_0^A f_y\ dA = 0$$

However, $f_y = f_c(y/c)$ as before, so

$$\frac{f_c}{c}\int_0^A y\ dA = 0$$

The quantity f_c/c is clearly not equal to zero; therefore,

$$\int_0^A y\ dA = 0$$

From Chapter 3 the reader should recall that this integral is the expression for the statical moment of the area in Figure A-3(b), using the neutral axis as the reference axis. The only way the statical moment of an area can be zero is if the reference axis is a centroidal axis.

The general formula for flexural stress developed here is subject to the following idealistic restrictions:

1. Transverse sections remain plane.
2. The beam is straight, of constant cross section, and does not twist under load.
3. The material is homogeneous and isotropic in the direction of stress.
4. The proportional limit is not exceeded.
5. The deformations are small.

DERIVATION OF BASIC
HORIZONTAL SHEARING
STRESS EQUATION

As with the derivation of the flexural stress formula, a rectangular cross section will be used here. This is done only for simplicity and the formula so developed is not restricted as to cross-sectional shape.

The beam shown with its moment diagram in Figure B-1(a) is typical in that its moment varies from one transverse section to another, e.g., from section 1 to section 2. In this case, M_2 is slightly greater than M_1, by the amount dM.

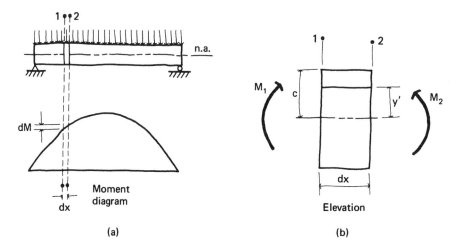

Figure B-1 Change in bending moment.

Figure B-2 shows this difference in moment between sections 1 and 2 by showing the normal stresses that act on those sections. There is an increase in these stresses acting on the transverse sections containing the element dA as we move from plane 1 to plane 2. This change in stress will result in an unbalanced force on the block that has dimensions b, $c - y'$, and dx. Therefore, a stress f_v is developed to put this block back in equilibrium. The magnitude of this horizontal shearing stress at level y can be determined from the basic equilibrium equation,

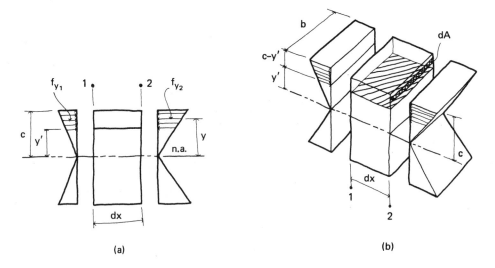

Figure B-2 Change in bending stress.

$$\sum F_x = 0$$
$$f_v b \, dx = C_2 - C_1 \tag{B-1}$$

where C_2 and C_1 of Figure B-3 are resultants or summations of the normal stresses, shown in Figure B-2(b) acting on the elemental areas dA.

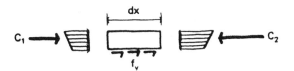

Figure B-3 Development of horizontal shearing stress.

$$C_2 = \int_{y_1}^{c} f_{y_2} dA$$

$$C_1 = \int_{y_1}^{c} f_{y_1} dA$$

In these equalities, f_y can be replaced by the appropriate values of My/I, where M is a constant for a given section and I is presumed constant for the entire beam.

$$C_2 = \frac{M_2}{I} \int_{y'}^{c} y \, dA$$

$$C_1 = \frac{M_1}{I} \int_{y'}^{c} y \, dA$$

Substituting into Equation (B-1), we get

$$f_v b \, dx = \frac{M_2 - M_1}{I} \int_{y'}^{c} y \, dA$$

where $M_2 - M_1 = dM$, as previously indicated. Solving for the shearing stress at level y yields

$$f_v = \frac{dM}{dx \, Ib} \int_{y'}^{c} y \, dA$$

where $dM/dx = V$, the vertical shearing force:

$$f_v = \frac{V}{Ib} \int_{y'}^{c} y \, dA$$

The expression $\int_{y}^{c} y \, dA$ has been given the symbol Q. It is the statical moment of that portion of the cross section which lies between level y' (where we are finding the stress) and level c (the edge of the section), taken with respect to the neutral axis. Using the symbol Q for this statical moment, the complete formula is

$$f_v = \frac{VQ}{Ib} \tag{8-1}$$

Because the flexure formula, $f_y = My/I$, was used in this derivation, the shearing stress formula is subject to the same assumptions or restrictions listed in Appendix A.

DERIVATION OF EULER
COLUMN BUCKLING EQUATION

Under the application of a certain critical load, a column can be in equilibrium in the curved (buckled) position. In this buckled position, an increase in P will cause an increase in the lateral deflection (leading to failure), and a decrease in P will cause the column to return to its initially straight position. It is this critical value of P that is quantified by the Euler equation.

Just as in a beam, the rate of change of slope of the column is directly proportional to the bending moment and inversely proportional to the stiffness. With respect to the free-body diagram in Figure C-1(b), this relationship is given by

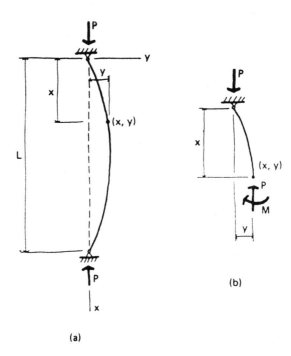

(a)

(b)

Figure C-1 Forces involved in column buckling.

$$-\frac{d^2y}{dx^2} = \frac{Py}{EI}$$

(The negative sign is present because of the selection of the origin for the coordinate axes.) As x increases, the slope decreases; thus, the rate of change of slope is negative. The equation can be rewritten as

$$\frac{d^2y}{dx^2} + \frac{P}{EI}y = 0 \qquad\qquad (C-1)$$

If we let $m = \sqrt{P/EI}$ such that

$$\frac{d^2y}{dx^2} + m^2 y = 0 \qquad\qquad (C-2)$$

the solution of the differential equation is of the form

$$y = A \cos mx + B \sin mx \qquad\qquad (C-3)$$

which involves two arbitrary constants, A and B. That this is a valid solution for y may be verified by taking the second derivative of Equation (C-3) and substituting it into Equation (C-2).

To evaluate the constants A and B, we can use the boundary condition of $y = 0$ when $x = 0$, from which we get $A = 0$, and therefore $y = B \sin mx$. Likewise, $y = 0$ when $x = L$ or $B \sin mL = 0$. If B is to have a value, then $\sin mL$ must be zero. This is only true if $mL = 0, \pi, 2\pi, 3\pi$, etc. The coefficient of π represents the buckling mode. For the single wave mode of our column, $mL = \pi$. Replacing m by $\sqrt{P/EI}$ and solving for P, we get

$$P = \frac{\pi^2 EI}{L^2} \qquad\qquad (10-2)$$

where P is the critical buckling load.

WEIGHTS OF SELECTED BUILDING MATERIALS*

	lb/ft³
Aluminum	160
Brick masonry construction	120
Cement plaster	120
Concrete masonry construction, hollow blocks	80
Concrete, stone, reinforced	150
Concrete, structural lightweight, reinforced	110
Earth, sand, loose	100
Earth, topsoil, packed	90
Glass	180
Gypsum board	50
Insulation, rigid	20
Plywood	40
Steel	490
Stone	160
Wood, Douglas fir	30
Wood, oak	45
Wood, redwood	25
Water, fresh	62

*Values in this table are intended to be representative rather than precise. Most material densities vary considerably, depending upon type and/or ambient conditions.

PROPERTIES OF SELECTED MATERIALS*

MATERIAL	STRENGTH (PSI) (YIELD VALUES EXCEPT WHERE NOTED)			MODULUS OF ELASTICITY (*E*) (KSI)	COEFFICIENT OF THERMAL EXPANSION (°F^{-1})(10^{-6})
	Tension	Compression	Shear		
Wood (dry)[†]					
Douglas fir	6 000	3 500	500	1 700	2
Redwood	6 500	4 500	450	1 300	2
Southern pine	8 500	5 000	600	1 700	3
Steel					
Mild, low-carbon	36 000	36 000	20 000	29 000	6.5
Cable, high-strength	275 000[‡]	—	—	25 000	6.5
Concrete					
Stone	200[‡]	3 500[‡]	180[‡]	3 500	5.5
Structural, lightweight	150[‡]	3 500[‡]	130[‡]	2 100	5.5
Brick masonry	300[‡]	4 500[‡]	300[‡]	4 500	3.4
Aluminum, structural	30 000	30 000	18 000	10 000	12.8
Iron, cast	20 000[‡]	85 000[‡]	25 000[‡]	25 000	6
Glass, plate	10 000[‡]	36 000[‡]	—	10 000	4.5
Polyester, glass-reinforced	10 000[‡]	25 000[‡]	25 000[‡]	1 000	35

*Values given in this table are intended to be representative rather than precise. Many properties vary, depending upon type, manufacturing process, and conditions of use.

[†]Values given are for the parallel-to-grain direction.

[‡]Denotes ultimate strength.

F

PROPERTIES OF AREAS

Table F-1 Areas and Centroids

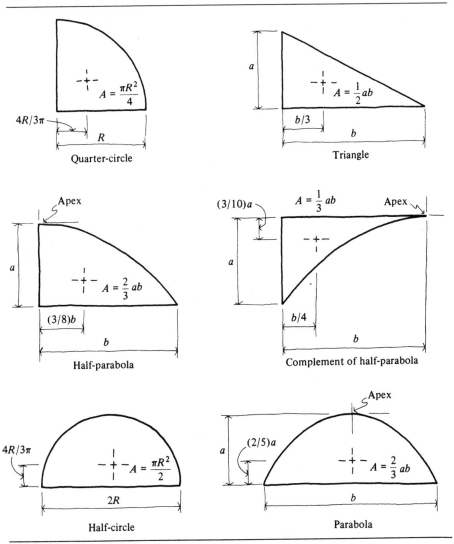

Quarter-circle

$A = \dfrac{\pi R^2}{4}$

$4R/3\pi$

R

Triangle

$A = \dfrac{1}{2}ab$

a

$b/3$

b

Half-parabola

Apex

$A = \dfrac{2}{3}ab$

a

$(3/8)b$

b

Complement of half-parabola

$(3/10)a$

$A = \dfrac{1}{3}ab$

Apex

a

$b/4$

b

Half-circle

$4R/3\pi$

$A = \dfrac{\pi R^2}{2}$

$2R$

Parabola

Apex

a

$(2/5)a$

$A = \dfrac{2}{3}ab$

b

Table F-2 Moments of Inertia

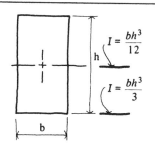

$I = \dfrac{bh^3}{12}$

$I = \dfrac{bh^3}{3}$

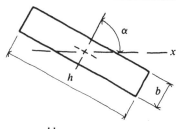

$I_x = \dfrac{bh}{12}(b^2 \sin^2 \alpha + h^2 \cos^2 \alpha)$

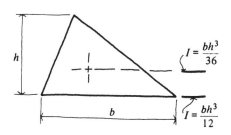

$I = \dfrac{bh^3}{36}$

$I = \dfrac{bh^3}{12}$

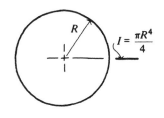

$I = \dfrac{\pi R^4}{4}$

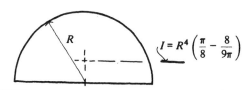

$I = R^4 \left(\dfrac{\pi}{8} - \dfrac{8}{9\pi} \right)$

PROOF OF MOMENT-AREA
THEOREMS

The radius of curvature at any point along a curve whose equation is $y = f(x)$ (Figure G-1) can be expressed as

$$R = \frac{[1 + (dy/dx)^2]^{3/2}}{d^2y/dx^2}$$

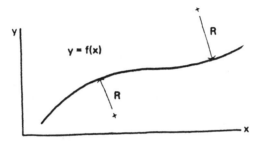

Figure G-1 General curve.

In beams, the slopes of elastic curves are always very small, and for this reason the term $(dy/dx)^2$ is insignificant and may be taken as zero. Therefore,

$$R = \frac{1}{d^2y/dx^2}$$

where d^2y/dx^2 is merely the rate of change of the slope. It is a measure of the change in slope between two points on the elastic curve as we shall use it. If the points a and b in Figure G-2 are allowed to approach each other, then θ_{ab} can be represented by d^2y/dx^2.

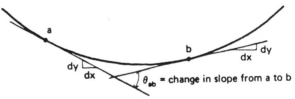

Figure G-2

If we equate the two values obtained for R,

$$R = \frac{EI}{M} = \frac{1}{d^2y/dx^2}$$

we get

$$EI\frac{d^2y}{dx^2} = M \qquad\qquad\text{(G-1)}$$

Credit for the development of this equation is given to Leonhard Euler (1707–1783), a Swiss mathematician.

First Moment-Area Theorem

The basic equation developed above can be written as

$$\frac{d(dy/dx)}{dx} = \frac{M}{EI}$$

For the small slopes of elastic curves assumed previously, $dy/dx = \tan\theta = \theta$. Then

$$\frac{d^2y}{dx^2} = \frac{d\theta}{dx}$$

Also,

$$\frac{d\theta}{dx} = \frac{M}{EI}$$

which can be written as

$$d\theta = \frac{M}{EI}dx$$

which is valid for a small length of beam dx. It states that a small change in angle is equal to a small area. Integrating both sides of the equation with finite limits determined by the portion of the beam shown in Figure G-3, we get that the change in angle θ from point A to point B is equal to the shaded area under the M/EI curve.

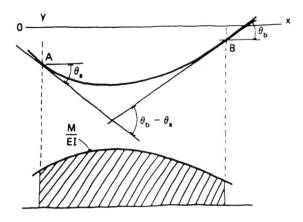

Figure G-3 Portion of elastic curve and *M/EI* diagram.

Second Moment-Area Theorem

Two tangent lines to the elastic curve at infinitesimally close points c and d will subtend a distance dt on the vertical line through B, as shown in Figure G-4.

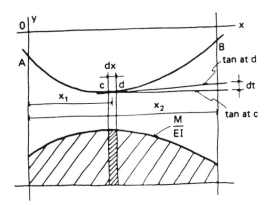

Figure G-4 Portion of elastic curve and *M/EI* diagram.

$$\tan d\theta = \frac{dt}{x_2 - x_1}$$

Since for small values of θ, the tangent function equals the angle itself, we can say that

$$d\theta = \frac{dt}{x_2 - x_1}$$

Previously, $d\theta$ was declared equal to $(M/EI)dx$; therefore,

$$dt = (x_2 - x_1)\frac{M}{EI}dx$$

This states that a small vertical distance on a line through B is equal to a small area, $(M/EI)dx$, times the distance from the line to the centroid of the area, $x_2 - x_1$. Integrating both sides of the equation with finite limits determined by the portion of the beam shown in Figure G-4, we get that the distance $t_{b/a}$ along a vertical line through B is equal to the shaded area times the distance from the vertical line to the centroid of the shaded area (see Figure G-5).

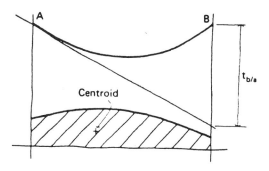

Figure G-5

ALLOWABLE STRESS VALUES FOR SELECTED WOODS

SPECIES AND SIZE	F_b (PSI)	F_b (PSI) (REPETITIVE)*	F_v (PSI)	F_c (PSI)	E ($\times 10^6$) (PSI)
Douglas fir No. 2					
thickness ≤ 3½	1250	1450	95	1050	1.7
> 3½	875	—	85	600	1.3
Eastern spruce No. 2					
thickness ≤ 3½	825	950	70	700	1.4
> 3½	575	—	65	375	1.0
Hem-fir No. 2					
thickness ≤ 3½	1000	1150	75	875	1.4
> 3½	675	—	70	475	1.1
Southern pine No. 2					
thickness ≤ 3½	1200	1400	90	1000	1.6
> 3½	1100	—	95	625	1.4
Eastern hemlock No. 2					
thickness ≤ 3½	1050	1250	85	900	1.1
> 3½	750	—	80	550	0.9

*Repetitive members in bending are ones that are closely enough spaced or tied sufficiently to avoid independent action, e.g., residential floor joists with bridging and subfloor.

When the duration of loading is known, the allowables may be modified by the following factors:

Two months (snow)	1.15
Seven days	1.25
Wind or earthquake	1.33
Impact	2.00

If the *full* load is to be applied, continuously or cumulatively, for more than 10 years, the allowable stresses should be multiplied by 0.90. The modulus of elasticity values are not subject to such modifications.

WOOD SECTION PROPERTIES

NOMINAL SIZE	ACTUAL SIZE (IN.)	AREA (IN.^{2})	S_x (IN.^{3})	I_x (IN.^{4})	I_y (IN.^{4})
2 × 4	1½ × 3½	5.25	3.06	5.36	0.984
2 × 6	1½ × 5½	8.25	7.56	20.8	1.55
2 × 8	1½ × 7¼	10.9	13.1	47.6	2.04
2 × 10	1½ × 9¼	13.9	21.4	98.9	2.60
2 × 12	1½ × 11¼	16.9	31.6	178	3.16
4 × 4	3½ × 3½	12.3	7.15	12.5	12.5
4 × 6	3½ × 5½	19.3	17.6	48.5	19.7
4 × 8	3½ × 7¼	25.4	30.7	111	25.9
4 × 10	3½ × 9¼	32.4	49.9	231	33.0
4 × 12	3½ × 11¼	39.4	73.8	415	40.2
4 × 14	3½ × 13¼	46.4	102	678	47.3
4 × 16	3½ × 15¼	53.4	136	1034	54.5
6 × 6	5½ × 5½	30.3	27.7	76.3	76.3
6 × 8	5½ × 7½	41.3	51.6	193	104
6 × 10	5½ × 9½	52.3	82.7	393	132
6 × 12	5½ × 11½	63.3	121	697	159
6 × 14	5½ × 13½	74.3	167	1128	187
6 × 16	5½ × 15½	85.3	220	1707	215
6 × 18	5½ × 17½	96.3	281	2456	243

J

PROPERTIES OF SELECTED STEEL SECTIONS

DESIGNATION	A (IN²)	d (IN.)	b_f (IN.)	S_x (IN³)	I_x (IN⁴)	I_y (IN⁴)
C 6 × 10.5	3.09	6.00	2.034	5.06	15.2	0.866
C 8 × 13.75	4.04	8.00	2.343	9.03	36.1	1.53
C10 × 25	7.35	10.00	2.886	18.2	91.2	3.36
C10 × 30	8.82	10.00	3.033	20.7	103	3.94
W 6 × 12	3.55	6.03	4.000	7.31	22.1	2.99
W 6 × 25	7.34	6.38	6.080	16.7	53.4	17.1
W 8 × 10	2.96	7.89	3.940	7.81	30.8	2.09
W 8 × 35	10.3	8.12	8.020	31.2	127	42.6
W 8 × 67	19.7	9.00	8.280	60.4	272	88.6
W10 × 12	3.54	9.87	3.960	10.9	53.8	2.18
W10 × 30	8.84	10.47	5.810	32.4	170	16.7
W10 × 49	14.4	9.98	10.000	54.6	272	93.4
W12 × 16	4.71	11.99	3.990	17.1	103	2.82
W12 × 50	14.7	12.19	8.080	64.7	394	56.3
W12 × 120	35.3	13.12	12.320	163	1070	345
W14 × 22	6.49	13.74	5.000	29.0	199	7.00
W14 × 48	14.1	13.79	8.030	70.3	485	51.4
W14 × 370	109	17.92	16.475	607	5440	1990
W16 × 31	9.12	15.88	5.525	47.2	375	12.4
W16 × 57	16.8	16.43	7.120	92.2	758	43.1
W18 × 40	11.8	17.90	6.015	68.4	612	19.1
W18 × 76	22.3	18.21	11.035	146	1330	152
W21 × 44	13.0	20.66	6.500	81.6	843	20.7
W21 × 50	14.7	20.83	6.530	94.5	984	24.9
W24 × 68	20.1	23.73	8.965	154	1830	70.4
W24 × 76	22.4	23.92	8.990	176	2100	82.5
W27 × 84	24.8	26.71	9.960	213	2850	106
W27 × 102	30.0	27.09	10.015	267	3620	139
W30 × 99	29.1	29.65	10.450	269	3990	128
W30 × 173	50.8	30.44	14.985	539	8200	598
W33 × 118	34.7	32.86	11.480	359	5900	187
W33 × 241	70.9	34.18	15.860	829	14200	932
W36 × 260	76.5	36.26	16.550	953	17300	1090
W36 × 300	88.3	36.74	16.655	1110	20300	1300

SHEAR, MOMENT, AND DEFLECTION EQUATIONS

①

$$V_{max} = P$$

$$M_{max} = PL$$

$$\Delta_{max} = \frac{PL^3}{3\,EI}$$

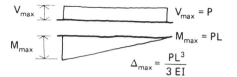

② $W = wL$

$$V_{max} = wL$$

$$M_{max} = \frac{wL^2}{2}$$

$$\Delta_{max} = \frac{wL^4}{8\,EI}$$

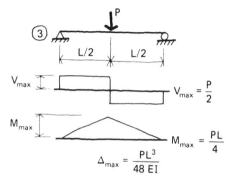

③

L/2   L/2

$$V_{max} = \frac{P}{2}$$

$$M_{max} = \frac{PL}{4}$$

$$\Delta_{max} = \frac{PL^3}{48\,EI}$$

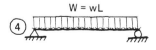

④ $W = wL$

$$V_{max} = \frac{wL}{2}$$

$$M_{max} = \frac{wL^2}{8}$$

$$\Delta_{max} = \frac{5\,wL^4}{384\,EI}$$

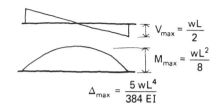

⑤ $(a < b)$

a   b

$$V_{max} = \frac{Pb}{L}$$

$$M_{max} = \frac{Pab}{L}$$

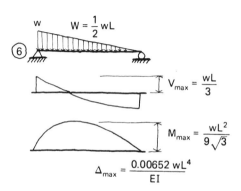

⑥ w $W = \frac{1}{2}wL$

$$V_{max} = \frac{wL}{3}$$

$$M_{max} = \frac{wL^2}{9\sqrt{3}}$$

$$\Delta_{max} = \frac{0.00652\,wL^4}{EI}$$

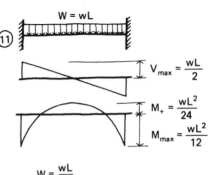

⑪

$$V_{max} = \frac{wL}{2}$$

$$M_+ = \frac{wL^2}{24}$$

$$M_{max} = \frac{wL^2}{12}$$

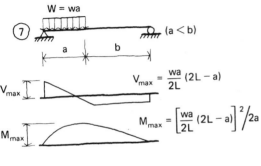

⑦ $(a < b)$

$$V_{max} = \frac{wa}{2L}(2L - a)$$

$$M_{max} = \left[\frac{wa}{2L}(2L - a)\right]^2 / 2a$$

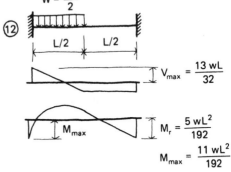

⑫

$$V_{max} = \frac{13\,wL}{32}$$

$$M_r = \frac{5\,wL^2}{192}$$

$$M_{max} = \frac{11\,wL^2}{192}$$

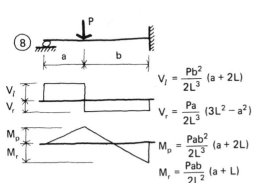

⑧

$$V_l = \frac{Pb^2}{2L^3}(a + 2L)$$

$$V_r = \frac{Pa}{2L^3}(3L^2 - a^2)$$

$$M_p = \frac{Pab^2}{2L^3}(a + 2L)$$

$$M_r = \frac{Pab}{2L^2}(a + L)$$

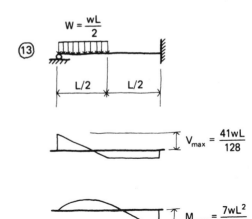

⑬

$$V_{max} = \frac{41wL}{128}$$

$$M_{max} = \frac{7wL^2}{128}$$

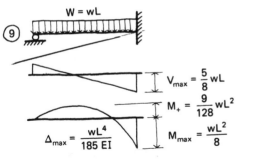

⑨

$$V_{max} = \frac{5}{8}wL$$

$$M_+ = \frac{9}{128}wL^2$$

$$\Delta_{max} = \frac{wL^4}{185\,EI}$$

$$M_{max} = \frac{wL^2}{8}$$

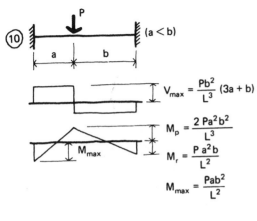

⑩ $(a < b)$

$$V_{max} = \frac{Pb^2}{L^3}(3a + b)$$

$$M_p = \frac{2\,Pa^2b^2}{L^3}$$

$$M_r = \frac{P\,a^2b}{L^2}$$

$$M_{max} = \frac{Pab^2}{L^2}$$

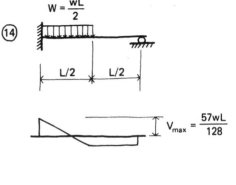

⑭

$$V_{max} = \frac{57wL}{128}$$

$$M_{max} = \frac{9wL^2}{128}$$

answers
to
problems

2-1.

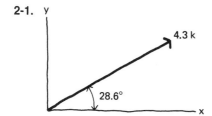

2-2.

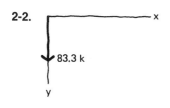

2-5.

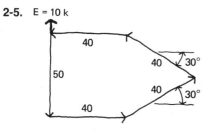

2-6. A = 50 k tension
B = 50 k tension

2-7. A = 24 k tension
B = 18 k compression

2-8. A = 198 k tension
B = 100 k tension

2-9. A and B approach infinity

2-10. M_O = 460 k-ft ↻
M_A = 255 k-ft ↻
M_B = 260 k-ft ↺

2-12. A_x = 0
A_y = 16.4 k ↑
B_y = 17.6 k ↑

2-13. A_y = 0
B_x = 0
B_y = 30 k ↑

2-14. A_y = 7.1 k ↑
B_x = 32 k ←
B_y = 7.1 k ↓

2-15. A_y = 54 k ↑
M_A = 270 k-ft ↺

2-16. (a) A_y = 75 k ↑
B_x = 0
B_y = 25 k ↓
   (b) A_x = 150 k →
A_y = 75 k ↑
B_x = 150 k ←
B_y = 25 k ↓

2-17. A_x = 60 k →
A_y = 30 k ↑
B_x = 60 k ←
B_y = 10 k ↑

2-18. A_x = 30 k →
A_y = 30 k ↑
B_x = 30 k ←

2-19. A_x = 30 k →
A_y = 15 k ↑
B_x = 30 k ←
B_y = 15 k ↑

2-20. A_x = 9 k ←
A_y = 6.75 k ↓
B_y = 6.75 k ↑

2-21. $A_x = 4$ k $\rightarrow$

$A_y = 4$ k $\downarrow$

$B_x = 4$ k $\leftarrow$

$B_y = 4$ k $\uparrow$

2-22. $A_x = 36$ k $\rightarrow$

$A_y = 9$ k $\uparrow$

$B_x = 36$ k $\leftarrow$

$B_y = 24$ k $\uparrow$

2-23. (a) stable, determinate
(b) stable, indeterminate, first degree
(c) stable, determinate
(d) stable, determinate
(e) unstable
(f) stable, indeterminate, second degree

2-25. Maximum sag $= 32$ ft
Maximum tension $= 9.6$ k

2-26. $T_{ab} = 56.6$ k
$C_{bc} = 80$ k

2-27. $P_1 \approx 13$ k
$P_2 \approx 47$ k

2-28. (a) $T = 155$ k
(b) $T = 168$ k
(c) $T = 212$ k
(d) $T = 404$ k
(e) $T = 765$ k

2-29. $T = 244$ k

2-30. $H_L = 360$ k
$x = 45$ ft

3-1. $\bar{y} = 6.5$ in

3-2. $\bar{x} = 3.33$ in
$\bar{y} = 3.33$ in

3-3. $\bar{y} = 6.5$ in

3-4. $\bar{x} = 28.6$ ft
$\bar{y} = 18.3$ ft

3-5. $\bar{y} = 20.1$ in

3-6. $\bar{y} = 0.7$ in

3-7. $I_{xx} = 981$ in^4
$I_{yy} = 469$ in^4

3-8. $I_{xx} = 122$ in^4
$I_{yy} = 72$ in^4

3-9. $I_{xx} = 1458$ in^4

3-10. $I_{xx} = 1976$ in^4

3-11. $I_{xx} = 5512$ in^4

3-12. $\bar{y} = 6.34$ in
$I_{xx} = 139.5$ in^4

3-13. $\bar{y}$ = 5.1 in
I_{xx} = 240 in^4

3-14. I_{xx} = 139.5 in^4

3-15. I_{xx} = 529.6 in^4

3-16. (a)   0.6%
(b)   35.4%

3-17. 5.4%

3-18. $\bar{y}$ = 13 in
I_{xx} = 25 500 in^4

3-19. r_x = 4.19 in
r_y = 0.96 in

3-20. r_x = 1.16 in

3-21. s = 3.06 in

4-1. f_a = 12.5 ksi

4-2. f_a = 9.55 ksi

4-3. ϵ = 0.000 69

4-4. δ = 1.2 ft

4-5. E = 3.77(10)^6 psi

4-6. f = 100 ksi

4-7. D = 4.1 in

4-8. δ = 0.15 in

4-9. L_1 = 399.84 ft
L_2 = 400.16 ft

4-10. width = 0.4 in

4-11. f = 33.9 ksi

4-12. L = 5.65 ft

6-1.

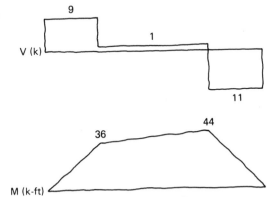

6-2.

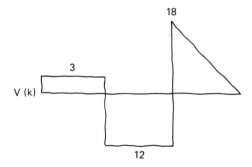

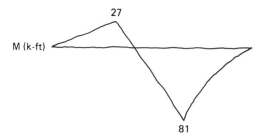

6-3.

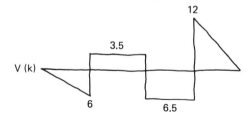

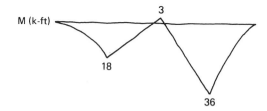

6-4.

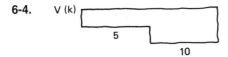

V (k)

5

10

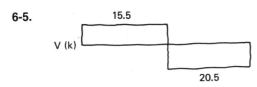

M (k-ft)

30

90

6-5.

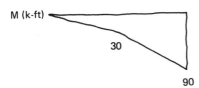

15.5

V (k)

20.5

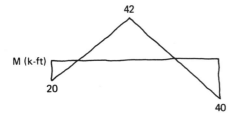

42

M (k-ft)

20

40

6-6.

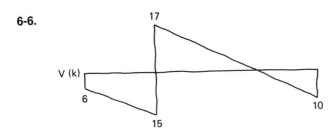

17

V (k)

6

15

10

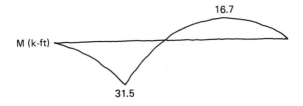

16.7

M (k-ft)

31.5

6-7.

22.5

V (k)

25.5

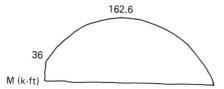

162.6

36

M (k-ft)

6-8.

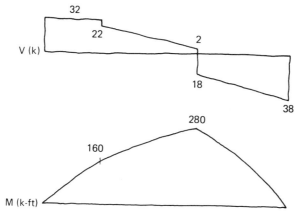

32

22

2

V (k)

18

38

280

160

M (k-ft)

6-9.

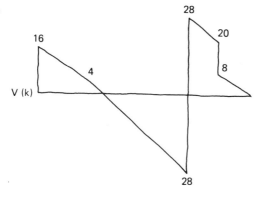

28

20

16

4

8

V (k)

28

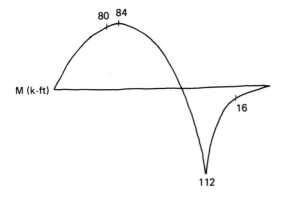

6-10.

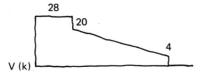

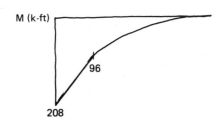

6-11.

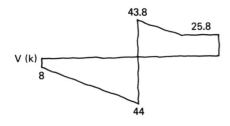

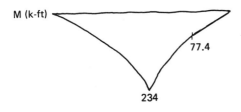

6-12.
$$M = \frac{wL^2}{12}$$

6-13.

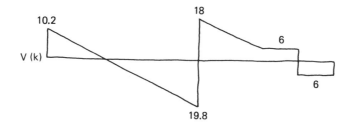

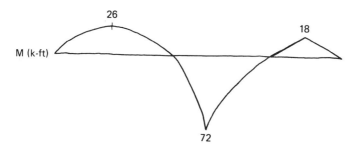

7-1. $f_b = 16.5$ ksi

7-2. $w = 106$ lb/ft

7-3. $L = 20.9$ ft

7-4. $f_{b_{\text{tens}}} = 1930$ psi

  $f_{b_{\text{comp}}} = 1610$ psi

7-5. (a) $f_b = 14.3$ ksi
      (b) $f_b = 1.08$ ksi

7-6. $f_b = 2070$ psi

7-7. $f_{b_{\text{comp}}} = 9.24$ ksi

  $f_{b_{\text{tens}}} = 22.4$ ksi

7-8. $f_b = 21.1$ ksi

7-9. (c) W36 × 170

7-10. (a) 2 × 12
      (b) 2 × 10
      (c) 2 × 10
      (d) 2 × 8

7-11. $L = 243.3$ ft

7-12. (a) $f_b = 122$ psi
      (b) $f_b = 840$ psi

8-1.

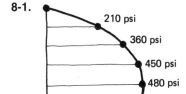

210 psi
360 psi
450 psi
480 psi
450 psi
90 psi

8-3. b = 3.3 in

8-4. Shearing stress maximizes just above the bottom of the notch.

8-5. 57.8 psi < 90 psi, OK in shear

8-6. f_v = 86.0 psi
max

8-7. f_v = 45.3 psi
max

8-8. 12.5%

8-9. f_v = 11.3 ksi

9-1. $\theta = \dfrac{wL^3}{6\,EI}, \Delta = \dfrac{wL^4}{8\,EI}$

9-2. 0.5 in < 0.8 in, Δ is OK

9-3. Δ_M = 2.31 in

9-4. Δ_{FE} = 2.36 in

9-5. Δ_{max} = 1.25 in

9-7.

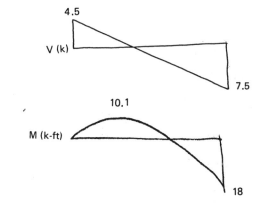

4.5

V (k)

7.5

10.1

M (k-ft)

18

9-8. $P_A = 16^K$, $P_B = 27^K$

9-9. $R_A = R_C = 0.3125\,P$, $R_B = 1.375\,P$

10.1. P_{cr} = 23.5 kips

10-2. $\left(\dfrac{P}{A}\right)_{cr} = 17.7$ ksi

10-3. $P_{cr} = 1370$ lb

10-4. $\left(\dfrac{P}{A}\right)_{cr} = 12.2$ ksi, $P_{cr} = 179$ kips

10-5. $P_{cr} = 40.3$ kips

10-6. $L = 26.6$ ft

10-7. $\left(\dfrac{P}{A}\right)_{cr} = 16.7$ ksi

10-8. $\left(\dfrac{P}{A}\right)_{cr} = 519$ psi, $P_{cr} = 4280$ lb

10-9. $X = 0.20$

11-1. AC = 30 kips C
CD = 30 kips C
DE = 112 kips C
EF = 97.5 kips C
FG = 0
EI = 0

BG = 30 kips C
AH = 112.5 kips T
HI = 135 kips T
IJ = 135 kips T
JB = 97.5 kips T
FJ = 37.5 kips T

AD = 116.7 kips C
HE = 31.8 kips C
EJ = 53.0 kips C
FB = 137.9 kips C
DH = 22.5 kips T

11-2. AC = BE = 40 kips C
AF = BF = 0
CF = EF = 33.6 kips T
CD = ED = 30.4 kips C
DF = 30 kips C

11-3. AC = 60 kips C
AE = DE = 0
CE = BE = 80 kips T
CD = BD = 100 kips C

11-4. AC = 67.5 kips C
AE = 25 kips T
CE = 70 kips T
EB = 90 kips T
CD = BD = 112.5 kips C
DE = 15 kips T

11-5. AB = 30 kips T
AC = BE = DE = 0
CD = 60 kips C
AD = 42.4 kips T
BD = 42.4 kips C

11-6. AC = BG = 40 kips C
AH = BJ = HI = JI = 120 kips T
CD = GF = DH = FJ = 0
AD = BF = 170 kips C
DI = FI = 56.6 kips T

DE = FE = 160 kips C

EI = 80 kips C

11-7. DE = 0

11-8. AB = 26.7 kips T

11-9. DB = 42.4 kips C

11-10. AB = 22.5 kips T

AC = 33.7 kips T

BE = 33.7 kips C

AD = 40.5 kips T

BD = 40.5 kips C

CD = 30 kips C

ED = 15 kips T

CF = 11.2 kips T

EH = 11.2 kips C

CG = 27 kips T

EG = 27 kips C

FG = 22.5 kips C

HG = 7.5 kips T

FI = HK = JK = 0

FJ = 13.5 kips T

HJ = 13.5 kips C

IJ = 15 kips C

index

A

Aalto, Alvar, 208
Accuracy of Computations, 17–18
Allowable Stress. *See* Stress
Answers to Problems, 264–75
Area
    first moment of, 73
    properties of (*table*), 251–52
    second moment of, 81
    statical moment of, 73
Average unit strain. *See* Strain
Axes, principal, 81
Axial stress. *See* Stress

B

Bartning, Otto, 116
Bay, structural, 198
Beams
    deflection, 180–88
        equations for, 260–61

lateral bracing, 160
lateral buckling, 159–62
types of, 11
Bending, inefficiency of, 7, 146
Bracing
    of beams, 160
    of columns, 210
Buckling
    of beams, 159–62
    of columns, 198–215
    diagonal compression, 166
    effect of compression, 7
    inelastic, 215
Building material weights (*table*), 249

C

Cables, 60
Center of gravity, 72
Centroid, 72–78
Centroid location (*table*), 251
Codes, 17

Cologne Cathedral, 100
Columns
   bracing of, 210–13
   buckling of, 198–215
   end conditions, 204–5
   failure modes, 199–200
   inelastic buckling, 215
   slenderness ratio, 202
Components, 21
Compression
   diagonal, 164
   stress, 7
Concrete, nature of, 118–19
Connections, symbols, 41–44
Counters, 220
Couple, 36
Creep, 121–22
Crown Hall, 161
Cuvier, Georges, 10

D

Dead loads, 6
Deflection
   formulas, 190–92
   limitations (*table*), 180
   table, 260–61
   theory, 180–88
Determinacy, 57–59
   of trusses, 220
Diagonal compression, 164

E

Eames, Charles, 222
Economy, 4
Elastic buckling, columns, 198–215
Elastic curve, 181
Equilibrant, 26
Equilibrium, 7
   concurrent forces, 28–32

   equations, 38
   moment, 38
   simple cables, 60
   single members, 42–49
   truss joints, 221–28
   two-force members, 50–55
Euler equation, 201–14
   derivation, 247–48
Euler, Leonhard, 200

F

First moment. *See* Area
Flexural stress. *See* Stress
Flexure formula, 148–50
   derivation, 240–43
Force
   components, 21
   definition, 21
   direction of, 21
   equilibrant, 26
   moment of, 34
   polygon, 25–27
   redundant, 59
   resultant, 21
   sense of, 21
Free-body diagram (FBD), 45
Furness, Frank, 100

H

Hooke, Robert, 103
Horizontal shear. *See* Stress

I

Indeterminacy, 59
Inertia, moment of, 80–93
Inflection point, 126

J

Joint equilibrium (trusses), 221–28

L

Lateral bracing
  beams, 160
  columns, 210
Lateral buckling, beams, 159–62
Laugier, Abbe, 3
LeCorbusier, 3
Live loads, 6
Loads, types, 6

M

Maisons Dom-ino, 3
Masonry, 121
Materials
  characteristics of, 114–22
  properties (*table*), 250
  weights (*table*), 249
Maybeck, Bernard, 222
Mechanics, 20
Mechanics of materials, 20
Method of joints, 221–28
Method of sections, 230–34
Mies van der Rohe, 161
Modulus of elasticity, 106
Moment, definition of, 33
Moment-area theorems, 182
  proof of, 253–56
Moment arm, 34
Moment diagrams, 124–40
  sign convention, 125–26
  table, 260–61
Moment of inertia, 80–93, 146
  formulas (*table*), 252
Munday, Richard, 116

N

Navier, Claude L. M. H., 106, 148
Neutral axis, 147–48
Notre-Dame du Raincy Church, 115

O

Otto, Frei, 208

P

Panel points, 220
Parallel axis theorem, 89
Parent, Antoine, 148
Perret, Auguste, 115
Point of Inflection, 126
Ponding, 180
Principal axes, 81
Problem answers, 264–75
Properties of areas (*table*), 251–52
Properties of materials (*table*), 250
Properties of sections
  steel (*table*), 259
  wood (*table*), 258

R

Radius of gyration, 95–96
Reinforced concrete, nature of, 118–19
Resultant, 21
Rigid body, 42, 44

S

Ste. Chapelle Church, 115
Saint-Venant, Barré de, 181

Second moment. *See* Area
Section modulus, 156–58
Shear diagrams, 124–40
    sign convention, 125
    table, 260–61
Slenderness ration, 202
Span/depth ration (*table*), 12–15
Stability, 4, 57–58
Statical moment, 73
Statics, 20, 45
    of cables, 60–65
    procedures, 68–69
Steel
    nature of, 120–21
    section properties (*table*), 259
Stiffness, 4, 105–6
Stirling, James, 222
Stirrups, 166
Strain
    average unit, 102
    flexural, 147–48
    shearing, 102
    thermal, 109–10
Strength of structure, 4
Stress
    allowable in steel, 157
    allowable in wood (*table*), 257
    axial, 98
    bending, 7
    compressive, 7
    in connections, 101–2
    critical buckling, 202
    definition, 98
    flexural, 146–58
        derivation of formula, 240–43
    horizontal shearing, 164–78
        derivation of formula, 244–46
    normal, 99, 102
    shearing, 7
    tangential, 99, 102
    tensile, 7

    thermal, 109–10
    yield, 104
Structural bay, 198
Structural design, 6
Structural planning, 5–6
Structural systems, 12–15
Structure
    in building, 11–17
    definition, 1
    in nature, 7–9
Superposition
    indeterminate structures, 193–96
    principle of, 189
Support conditions
    fixed, 41
    hanger, 40
    pin, 40
    roller, 41

T

Tangential deviation, 182
Temperature, effects of, 108–10
Tension
    diagonal, 164. *See also* Stress
Truss
    analysis, 221–34
        method of joints, 221–28
        method of sections, 230–34
    definition, 218
    open-web bar joist, 236
    types, 219, 235–37
    Vierendeel, 236
Two-force member, 50–55, 221

V

Varignon, Pierre, 35
Varignon's theorem, 35

Vierendeel, Arthur, 236
Vierendeel ''truss'', 236–37

W

Weights, building materials (*table*), 249
Wood
    allowable stresses (*table*), 257
    nature of, 114–18
    section properties (*table*), 258
Wright, Frank Lloyd, 222

Y

Yield point, 103–4
Young, Thomas, 106
Young's modulus, 106

Z

Zero members, in trusses, 221
Zero shear, in beams, 135–36